D1074987

THE MARS MYSTERY

Also by Graham Hancock

Journey Through Pakistan
Ethiopia: The Challenge of Hunger
AIDS: The Deadly Epidemic
Lords of Poverty
African Ark: Peoples of the Horn
The Sign and the Seal: A Quest for the Lost Ark of the Covenant
Fingerprints of the Gods
The Message of the Sphinx (with Robert Bauval)

THE MARS MYSTERY

The Secret Connection Between Earth and the Red Planet

GRAHAM HANCOCK

Crown Publishers, Inc. • New York

The excerpt from the poem "High Flight" by John Gillespie Magee, Jr., on page 79 is reprinted with permission from the International Herald Tribune.

Published by Crown Publishers, Inc., 201 East 50th Street, New York, New York 10022. Member of the Crown Publishing Group.

Random House, Inc. New York, Toronto, London, Sydney, Auckland
www.randomhouse.com

CROWN and colophon are trademarks of Crown Publishers, Inc.

Printed in the United States of America

Design by Lenny Henderson

Library of Congress Cataloging-in-Publication Data
Hancock, Graham.
The mars mystery : the secret connection between earth and the red planet / Graham Hancock.
p. cm.
1. Mars (Planet)—Surface. 2. Mars (Planet)—Geology. 3. Great Sphinx (Egypt). 4. Egyptology. 5. Life on other planets.
I. Title.
QB641.H317 1998
999'.23—dc21 98-17933
CIP

ISBN 0-609-60086-9

10 9 8 7 6 5 4 3 2 1

First Edition

Contents

AUTHOR'S NOTE

The Mars Mystery is being published in the U.S. under my sole name, since I have been the main author and coordinator. Nevertheless, I feel it is important for readers to know that the book *is* a work of coauthorship. To be specific, I was the sole author of chapters 1 through 4 and chapters 18 through 26. My research assistant John Grigsby wrote chapters 5 through 16 (with contributions from Robert Bauval in chapter 16). Chapter 17 was largely written by me with contributions from John Grigsby and Robert Bauval.

Because of the collective nature of this work I have chosen to adopt the "we" tone of voice throughout the story. When references are made to "our" previous publications I am speaking primarily of my book *Fingerprints of the Gods,* of Robert Bauval's book *The Orion Mystery,* and of the book that Robert and I wrote together, *The Message of the Sphinx.*

Thanks to Chris O'Kane of the Mars Project U.K., and to Simon Cox, for library and documentation research on our behalf. Special thanks also to Dr. Benny Peiser of Liverpool's John Moore's University, who kindly put his personal library at our disposal.

I would like to add that a major part of the function of *The Mars Mystery* is to draw public attention to discoveries made by scientists around the world concerning the Mars anomalies and concerning the extremely grave and pressing issue of planetary cataclysms. Without the dedicated, groundbreaking work of these scientists, there would have been no book for us to write. We have attempted to report and represent their work fairly, wherever possible in their own words, but the overall conclusions that we have drawn are our own. Our role in this respect has been as synthesizers, connecting evidence and data from many different fields of research. It was only as we began to put the pieces of the jigsaw puzzle together that we ourselves became aware of the big picture and of the truly alarming implications that it has not only for the past of Earth but also for its future.

Graham Hancock

PART ONE

The Murdered Planet

1

Parallel World

ALTHOUGH separated by tens of millions of miles of empty space, Mars and Earth participate in a mysterious communion.

Repeated exchanges of materials have taken place between the two planets—the most recent involving spacecraft from Earth that have landed on Mars. Likewise we now know that chunks of rock thrown off from the surface of Mars periodically crash into Earth. By 1997 a dozen meteorites had been firmly identified as having originated on Mars. They are known technically as SNC meteorites (after Shergotty, Nakhla, and Chassingy, the names given to the first three such meteorites found[1]) and researchers around the world are on the lookout for more.[2] According to calculations by Dr. Colin Pillinger of the U.K. Planetary Sciences Research Institute, "100 tons of Martian material arrives on Earth each year."[3]

One of the Mars meteorites, ALH84001, was found in Antarctica in 1984. It contains tiny tubular structures that NASA scientists sensationally identified in August 1996 as "possible microscopic fossils of bacteria-like organisms that may have lived on Mars more than 3.6 billion years ago."[4] In October 1996 scientists at Britain's Open University announced that a second Martian meteorite, EETA 79001, had also been found to contain the chemical signatures of life—in this case, astonishingly,

> organisms that could have existed on Mars as recently as 600,000 years ago.[5]

LIFE-SEED

Two probes were launched by NASA in 1996—*Pathfinder,* a lander-rover, and *Mars Global Surveyor,* an orbiter. Further missions are budgeted to follow through 2005, when an attempt will be made to scoop up a chunk of the surface rock or soil of Mars and then return the sample to Earth.[6] Russia and Japan are also sending probes to Mars to undertake a range of scientific tests and experiments.

Longer term are plans to "terraform" the Red Planet. This would involve the introduction of greenhouse gases and simple bacteria from Earth. Over a period of centuries the warming effects of the gases and the metabolic processes of the bacteria would transform the Martian atmosphere, making it habitable by more and more complex species—either introduced or locally evolved.[7]

How likely is it that humanity will be able to fulfill this plan to "seed" Mars with life?

Apparently it is only a matter of finding the money. The technology to do the job already exists.[8] Ironically, however, the existence of life on Earth itself remains one of the great unsolved mysteries of science. Nobody knows when, why, or how it began here. It just seems to have exploded suddenly, out of nowhere, at a very early stage in the planet's history. Although Earth is thought to have formed 4.5 billion years ago, the most ancient surviving rocks are younger than that—about 4 billion years old. Traces of microscopic organisms have been found going back almost 3.9 billion years.[9]

The transformation of inanimate matter into life is a miracle that has never repeated itself, one that even the most advanced scientific laboratories cannot replicate. Are we really to believe that such an amazing piece of cosmic alchemy could have occurred *by chance* in just the first few hundred million years of Earth's long existence?

SOME OPTIONS

Professor Fred Hoyle of Cambridge University does not think so. His explanation for the origin of life on Earth so soon after the formation of the planet is that it was imported from outside the solar system on great interstellar comets. Some fragments collided with Earth, releasing spores that had been held in suspended animation in the cometary ice. The spores spread out and took root all around the newly formed planet, which was soon densely colonized by hardy microorganisms. These slowly evolved and diversified, eventually producing the immense range of life-forms that we know today.[10]

An alternative and more radical theory, supported by a number of scientists, is that Earth could have been deliberately terraformed 3.9 billion years ago, just as we are now preparing to terraform Mars. This theory presupposes the existence of an advanced star-faring civilization—or more likely, many such civilizations—distributed throughout the universe.

Most scientists do not see the need for comets or aliens. Their theory, the mainstream view, is that life arose on Earth accidentally, without any outside interference. They further argue, on the basis of widely agreed calculations about the size and composition of the universe, that there are probably hundreds of millions of Earth-like planets spread randomly across billions of light years of interstellar space. They point out that it is improbable, amid such legions of suitable planets, that life would have evolved only on Earth.

WHY NOT MARS?

In our own solar system, the first planet out from the Sun, tiny, seething Mercury, is believed to be incongenial to any imaginable form of life. So too is Venus, the second planet from the Sun, where concentrated sulphuric acid pours down twenty-four hours a day from poisonous clouds. Earth is the third planet from the Sun. The fourth, Mars, is indisputably the most Earth-like in the solar system. Its axis is tilted at an angle of 24.935 degrees in relation to the plane of its orbit around the Sun (Earth's axis is tilted 23.5 degrees). It makes a complete rotation around its axis in 24 hours, 39 minutes, 36 seconds (Earth's rotational period is 23 hours, 56 minutes, 5 seconds). Like Earth, Mars is subject to the cyclic axial wobble that astronomers call precession. Like Earth it is not a perfect sphere but somewhat flattened at the poles and expanded into a bulge at the equator. Like Earth it has four seasons. Like Earth it has icy polar caps, mountains, deserts, and dust storms. And although Mars today is a freezing hell, there is evidence that in some ancient period it was alive with oceans and rivers and enjoyed a climate and atmosphere quite similar to those of Earth.

How probable is it that the spark that ignited life on Earth would not also have made its mark on neighboring, similar Mars? Whether Earth was deliberately terraformed, in other words, or whether it was seeded with the spores of life from crashed comets—or whether, indeed, life arose here spontaneously and accidentally—it is reasonable to hope that we might find traces of the same kind of process on Mars.

If such traces are not forthcoming, then the chances that we are alone in the universe increase and the chances of life being discovered anywhere else are dramatically reduced. The implication will be that Earth's life-forms emerged under conditions so focused, specialized, and unique—and at the same time so random—that they could not be replicated even on a nearby world belonging to the same solar family. How much less likely, therefore, that they could be replicated on alien worlds in orbit around distant stars.

For this reason the question of life on Mars must be regarded as one of the great philosophical mysteries of our time. With the rapid advances in exploration of the planet it is a mystery that is soon likely to be solved.

HINTS OF LIFE

The evidence in from Mars so far takes four principal forms:

1. Earth-based observations from telescopes
2. Observations and photographs from orbiting spacecraft
3. Chemical and radiological tests carried out on Martian soil samples by NASA landers, with the results being transmitted back to Earth for analysis
4. Microscopic examination of meteorites known to have come from Mars

In the late nineteenth and early twentieth centuries, Earth-based telescopes produced the first ever "life on Mars" sensation—the claim that the planet was checkered with a gigantic network of irrigation canals bringing water from the poles to the parched equatorial regions. This claim, which we shall discuss further in part 2 of this book, was put forward by Percival Lowell, a prominent U.S. astronomer, and made an indelible mark on the collective psyche of Americans. Most scientists ridiculed Lowell's ideas, however, and in the 1970s, NASA's *Mariner 9* and *Viking 1* and *2* probes orbited the planet and sent back definitive photographs proving that there were no canals.

It is now recognized that Lowell (and others who claimed to have seen the canals) were the victims of poor-quality telescopic images and an optical illusion that causes the brain to link disparate, unconnected features into straight lines. Even today, no Earth-based telescope has sufficient resolution to allow us to solve the mystery of life on Mars. We must therefore make our deductions using the three other types of evidence available to us—Martian meteorites, orbiter observations, lander observations.

We have already seen that two of the Martian meteorites appear to contain traces of primitive microorganisms (although many scientists disagree with this interpretation). Less well known is the fact that a number of the tests carried out in 1976 by the *Viking* landers also proved positive for life. The impression conveyed in public statements made at the time by NASA is that the planet is barren—because no organic molecules were found on the surface at either of the two landing sites. But puzzlingly, the Martian samples did give positive results for metabolic processes such as photosynthesis and chemosynthesis that are normally associated with life.[11] What is known as a gas-exchange experiment also produced a positive result with soil samples liberating substantial quantities of oxygen in response to treatment with an organic nutrient.[12] Another positive result produced in a "labeled-release" experiment was absent in a control sample that had been baked at a high temperature—precisely as one would expect if the original reaction had been caused by a biological agent.[13]

This leaves the orbiter observations. In frames sent back by *Mariner 9* and *Viking 1,* strangely familiar objects can be seen that have been interpreted by some scientists not only as signs of life but as evidence that advanced *intelligent* life must once have been present on Mars.

THE PYRAMIDS OF ELYSIUM

The earliest anomalous images were acquired during 1972 and show an area of Mars known as the Elysium Quadrangle. At first little attention was paid to these images. Then in 1974 a brief notice appeared in the scientific journal *Icarus.* Written by Mack Gipson, Jr., and Victor K. Ablordeppy, the article reported:

> Triangular and pyramid-like structures have been observed on the Martian surface. Located in the east central portion of the Elysium Quadrangle, these features are visible on the *Mariner* photographs, B frames MTVS 4205-3 DAS 07794853 and MTVS 4296-24 DAS 12985882. The structures cast triangular and polygonal shadows. Steep-sided volcanic cones and impact craters occur only a few kilometers away. The mean diameter of the triangular pyramidial structures at the base is approximately three kilometers and the mean diameter of the polygonal structures is approximately six kilometers.[14]

Another *Mariner* photograph, frame 4205-78, quite distinctly shows four massive three-sided pyramids. These were commented on in 1977 by the Cornell University astronomer Carl Sagan. "The largest," he wrote, "are three kilometers across at the base and one kilometer high—much larger than the pyramids of Sumer, Egypt, or Mexico on Earth. They seem to be eroded and ancient and are, perhaps, only small mountains, sandblasted for ages. But they warrant, I think, a careful look."[15]

What is particularly notable about the four structures captured in this latter frame is that they appear to have been set out on the Martian surface in a definite pattern or alignment very like pyramids on terrestrial sites. In this they also have much in common with other Martian "pyramids" that lie in a region known as Cydonia, at approximately 40 degrees north latitude, almost halfway around the planet from Elysium.

THE PYRAMIDS AND THE "FACE" OF CYDONIA

The Cydonia pyramids were photographed in 1976 by the *Viking 1* orbiter from an altitude of about 1,000 miles and were first identified on *Viking* frame 35A72 by Dr. Tobias Owen (now professor of astronomy at the University of Hawaii). The same frame, covering approximately 34 by 31 miles—about the size of Greater London—also shows many other features that could be artificial.

A casual glance reveals only a jumble of hills, craters, and escarpments. Gradually, however, as though a veil is being lifted, the blurred scene begins to feel organized and structured—too *intelligent* to be the result of random natural processes. Although the scale is grander, it does look the way some archaeological sites on Earth might look if photographed from 1,000 miles up. The more closely you examine the frame, the more it becomes apparent that it really could be an ensemble of enormous ruined monuments on the surface of Mars.

Of these by far the most dramatic is a gigantic Sphinx-like face that NASA officially dismisses as a trick of light and shadow.[16] This explanation began to be challenged seriously only after 1980, as we shall see in part 2, when Vincent DiPietro, himself a computer scientist with NASA's Goddard Spaceflight Center, discovered another image of the "Face" on frame 70A13. This second image, which had been acquired 35 Martian days later than the first one and under different lighting conditions, made possible comparative views and detailed measurements of the Face. Complete with its distinctive headdress, it is now known to be almost 1.6 miles in length from crown to chin, 1.2 miles wide, and just under 2,600 feet high.[17]

The Face could be a small mountain, naturally weathered. But how many mountains have left and right sides so intricately similar? Image analysts say that the "bilateral symmetry" of the Face, mimicking a natural, almost human appearance, is most unlikely to have come about by chance. And this impression is confirmed by other characteristics that have subsequently been identified under computer enhancement. These include "teeth" in the mouth, bilaterally crossed lines above the eyes, and regular lateral stripes on the headpiece—suggestive, to some researchers at least, of the *nemes* headdress of ancient Egyptian pharaohs.[18]

According to Dr. Mark Carlotto, an expert in image processing, "These features appear in *both* of the *Viking* images, are coherent shapes, and are structurally integral to the object; therefore they could not have been caused by random noise or by artifacts of the image restoration and enhancement process."[19]

"AN IMPROBABLE ASSORTMENT OF ANOMALIES . . ."

The same is also true for the D&M Pyramid (named after DiPietro and his associate Gregory Molenaar, who discovered it). This five-sided structure stands about ten miles from the Face and, like the Great Pyramid of Egypt, is aligned almost perfectly north-south—toward the spin axis of the planet. Its shortest side is a mile, its long axis extends to almost two miles, it is almost half a mile high, and it has been estimated to contain over a cubic mile of material.[20]

Commenting on the proximity of the Face and the D&M Pyramid, former NASA consultant Richard Hoagland asks a pointed question: "What are the odds against two 'terrestrial-like monuments' on such an alien planet and in essentially the same location?"[21]

Hoagland has made his own detailed study of frames 35A72 and 70A13 and has identified additional possibly artificial features. These include the so-called Fort, with its two distinctive straight edges, and the City, which he describes as "a remarkably rectilinear arrangement of massive structures interspersed with several smaller 'pyramids' (some at exact right angles to the larger structures) and even smaller conical-shaped 'buildings.' "[22] Hoagland also points out another striking fact about the City: it seems to have been purposefully sited in such a way that hypothetical inhabitants would have enjoyed a perfect, indeed almost ceremonial, view of the Face.[23]

The impression of a great ritual center, shrouded under the dust of ages, is enhanced by other features of Cydonia, such as the Tholus, a massive mound similar to Britain's Silbury Hill, and the City Square, a grouping of four mounds centered on a fifth, smaller mound. This configuration—suggestive of crosshairs—turns out to be located at the exact lateral center of the City.[24]

In addition, a group of British researchers based in Glasgow have recently identified what looks like a massive four-sided pyramid, the so-called NK Pyramid, 25 miles west of the Face and on the same latitude (40.8 degrees north) as the D&M Pyramid. "Looking at the whole of Cydonia and at the way all these structures are sited," says Chris O'Kane of the U.K. Mars Project, "my gut feeling is that they have to be artificial. I don't see any way that such a complex system of alignments could have come about by chance."[25]

O'Kane's hunch is strengthened by the fact that "many of the structures are non-fractal." In plain English this means that their contours have been scanned and assessed as artificial (rather than natural) by highly sophisticated computers of the type normally used in modern warfare to pinpoint the locations of camouflaged tanks and artillery in aerial reconnaissance photographs.

"What we have, therefore," sums up Chris O'Kane, "is an improbable assortment of anomalies. They have what look like planned alignments, they're found in distinctive groups, and they're non-fractal. All in all, we have to say this is highly unusual."[26]

Nor are Cydonia and Elysium the only sites to have yielded photographic evidence of unusual and apparently artificial structures. Other Martian features that are decidedly non-fractal include a straight line more than three miles long defined by a row of small pyramids; a single pyramid poised on the edge of a gigantic crater; extensive rhomboidal enclosures in the south polar region; and a weird, castle-like edifice rising to a steeple more than 2,000 feet high.[27]

GALLERY OF MYSTERIES

In 1996, during the last year of his life, Carl Sagan made a curious comment about the Face on Mars. This structure, he said, was "probably sculpted by slow geological processes over millions of years." Nevertheless he added:

> I could be wrong. It's hard to be sure about a world we've seen so little of in extreme close-up.[28]

Sagan urged that forthcoming American and Russian missions to Mars should make special efforts "to look much more closely at the pyramids and at what some people call the Face and the City. . . . These features merit closer attention with higher resolution. More detailed photos of the Face would surely settle issues of symmetry and help resolve the debate between geology and monumental structure."[29]

We do not share Sagan's confidence that high-resolution photographs will resolve the debate. Until astronauts land on Mars and explore Cydonia, even the best photographic images are likely to leave room for doubt—in both directions. Matters are further complicated by the fact that NASA's policy statements concerning the pyramids and the Face have frequently been bizarre and contradictory. Smacking of a secretive or even dishonest agenda, these statements have inevitably provoked some observers to make mental links between the "monuments" of Mars and the UFO controversy (Roswell, Area 51, alleged abductions by aliens, etc.). The effect has been to fuel the paranoia—particularly rampant in the United States—that a massive government cover-up is under way.

We will return to the pyramids and the Face of Mars in part 2 and investigate the allegations of conspiracy in part 3. Our immediate aim in part 1 is to explore the planet itself and to enter its gallery of mysteries.

The greatest mystery of all is why Mars died.

Is There Life on Mars?

An astronomer received the following telegram from a newspaper editor: WIRE ONE HUNDRED WORDS COLLECT. IS THERE LIFE ON MARS? The astronomer wired back, NOBODY KNOWS, repeated fifty times.[1]

That happened before the era of space exploration. Then, in July 1965, NASA's first successful probe—*Mariner 4*—was maneuvered into a fly-by of Mars and sent back 22 black-and-white television pictures showing the mysterious planet to be formidably cratered and, apparently, as completely lifeless as the Moon. In subsequent years *Mariner 6* and *7* also flew past Mars, and *Mariner 9* orbited it, sending back 7,329 pictures (1971–1972). In 1976 *Viking 1* and *2* went into long-term orbits during which they sent back more than 60,000 high-quality images and placed lander modules on the surface. Three Soviet probes also investigated Mars, two of them reaching its surface.[2]

Up until early 1998, the question "Is there life on Mars?" could still only be answered, "Nobody knows." With more data at their disposal, however, scientists have formed a range of opinions on the matter. Despite the planet's devastated appearance, many now agree that extremely simple bacteria-like or virus-like microorganisms could have survived beneath the surface. Others feel there is no life at all there now, but do not rule out the possibility that Mars could have had a "flourishing biota" in some distant past epoch.

A key element in the widening scientific debate, as we saw in chapter 1, is that a number of possible microfossils and chemical evidence for life processes have been detected in chunks of rock from Mars that have reached Earth as meteorites. This evidence must be set alongside the positive tests for life processes, also reviewed in chapter 1, that were carried out by the *Viking* landers.

TESTING POSITIVE

The story of the search for life on Mars has many puzzling elements. Among these is NASA's published official conclusion that the 1976 *Viking* mission

> found no persuasive evidence for life on the surface of the planet.[3]

Dr. Gilbert Levin, one of the principal scientists involved in *Viking*, cannot accept this. He carried out the labeled-release experiment described in chapter 1, which produced an unmistakably positive reading. He wished to announce it as such at the time, but other colleagues at NASA overruled him. "A number of explanations have been proposed to explain the results of my experiment," commented Dr. Levin in 1996. "None of them are convincing. I believe that Mars has life today."[4]

It appears that Levin was overruled because his test contradicted negative results in other tests that had been devised by more senior colleagues—thus potentially calling the judgment of those colleagues into question. Particular weight was put on the fact that *Viking*'s mass spectrometer had detected no organic molecules on Mars. Yet Levin has subsequently shown that the probe was equipped with a badly underpowered mass spectrometer. It had a minimum sensitivity of ten million biological cells in a sample, compared with sensitivities down to just fifty cells that can be achieved by other instruments.[5]

Levin was encouraged to speak out only after NASA's announcement in August 1996 that apparent traces of microfossils had been found in meteorite ALH84001. This evidence strongly supports Levin's own view that there has been life on the Red Planet all along, despite the extremely harsh conditions that prevail there.

> Life is hardier than we had ever imagined. Microbes have been found in nuclear fuel rods inside reactors and in the depths of the ocean where there is no light.[6]

Colin Pillinger, professor of planetary science at the U.K.'s Open University, agrees: "I passionately believe that conditions on Mars were once conducive to life," he says. He too points out that certain life-forms can survive in the most inimical conditions: "Some can hibernate at temperatures well below zero and there is tentative evidence for life at 150 Celsius. How much more tenacious can you get?"[7]

LIVING IN EXTREMES

Mars is bitterly cold, with an average temperature across the planet of minus 23 C, plummeting to minus 137 C in some locations.[8] There is an acute shortage of life-giving gases such as nitrogen and oxygen.[9] In addition, atmospheric pressure is low. A person standing at "Mars datum," an agreed elevation selected by scientists to serve as the equivalent of sea level on Earth, would experience an atmospheric pressure no stronger than the pressure exerted on Earth at 18 miles above sea level.[10] Under these low pressures and temperatures there is and can be no liquid water on Mars.

Scientists do not believe it is possible for life to emerge anywhere without the presence of liquid water. If this is true then evidence of past or present life on Mars must strongly imply that the planet was once endowed with large quantities of liquid water—something, as we shall see, for which there is overwhelming evidence. That the water has since been lost is not in doubt. However, this does not necessarily mean that no life could have survived. On the contrary, a number of recent scientific discoveries and experiments have demonstrated that, on Earth at least, life can flourish in just about any conditions.

In 1996 British scientists drilling more than 13,000 feet below the surface of the Atlantic Ocean, found "a thriving subterranean world of microscopic creatures. . . . [These] bacteria show it is possible for life to survive under extreme conditions where pressures are 400 times greater than at sea level and where temperatures can reach 170 degrees centigrade."[11]

Other researchers exploring active submarine volcanoes at depths of more than two miles have found animals from a phylum called Pogonophora grazing on colonies of bacteria that thrive in seething, mineral-rich plumes rising from the seabed. Normally only a few millimeters long, these wormlike creatures are here freakishly enlarged to huge sizes and seem to be mimicking the mythical salamander that was supposed to live in fire.

The bacteria on which Pogonophora feast are almost equally outlandish. They do not rely on sunlight for energy, since none filters down to these depths, but use "the heat of near-boiling water bubbling up from below the crust." They do not require organic detritus for nourishment but consume "minerals in the hot brines."[12] Referred to by zoologists under the general category of "extremophiles," such creatures include autotrophs that eat basalt, use hydrogen gas for energy, and extract carbon from inorganic carbon dioxide.[13] Other autotrophs

> have been found three kilometers below the surface, where the only source of heat is the heat of the rocks. . . . They have been found at temperatures of 113 C. . . . They have been found . . . in streams of acid; in toluene, benzene, cyclohexane, and kerosene; and at 11,000 meters down in the Marianas Trench.[14]

Creatures of this kind might conceivably have survived on Mars, perhaps locked in the 10-meter deep layer of permafrost that is believed to underlie the planet's surface,[15] perhaps in suspended animation, for immense periods of time. On Earth, dormant microbes inside insects preserved in amber for tens of millions of years were successfully revived by scientists in California in 1995 and placed in a quarantined lab.[16] Other viable microorganisms that have been isolated from salt crystals are more than 200 million years old.[17] In laboratory experiments: "Bacterial spores have been heated to boiling point and cooled to –270 degrees C, which is the temperature of space between the stars. When things get better they come to life again."[18] Likewise there are viruses that "can be activated in cells even if they are inert outside such bio-organization." In their inert state these frightening little entities—smaller than the wavelength of visible light—are almost literally immortal. On examination they are "extremely complicated having a genome composed of 1.5×10^4 nucleotides."[19]

As NASA continues its exploration of Mars, scientists believe that there is a very real possibility of cross-contamination. Indeed, cross-contamination could have occurred long before the epoch of spaceflight. Just as meteorites from the surface of Mars have reached Earth, it is considered highly probable that rocks "splashed off" Earth's surface by asteroid impacts have reached Mars. It is conceivable that the spores of life itself could have been carried to Earth on meteorites from Mars—or, vice versa, that the spores of life could have been carried from Earth to Mars. Paul Davis, professor of natural philosophy at Adelaide University, points out that "Mars is not an especially hospitable planet for terrestrial-type life. . . . Nevertheless, some species of bacteria found on Earth might be able to survive there. . . . If life had become firmly established on Mars in the remote past it could have gradually adapted to the present harsher environment as conditions slowly deteriorated."[20]

HIGH-STAKES DEBATE

Perhaps by coincidence, NASA chose a time when the implications of the survival of microorganisms in extreme environments were being widely discussed by scientists and in the media to announce the discovery of microfossils in meteorite ALH84001. According to Dr. David McKay, who led the team investigating the meteorite:

> There is not any one finding that leads us to believe that [there was] past life on Mars. Rather it is a combination of many things that we have found. . . . [These] include an apparently unique pattern of organic molecules, carbon compounds that are the basis of life. We also found several unusual mineral phases that are known products of primitive microorganisms on Earth. Structures that could be microscopic fossils seem to support this. The relationship of all of these things in terms of localization—within a few hundred thousands of an inch of one another—is the most compelling evidence.[21]

Many scientists do not find McKay's evidence so compelling. Among those who disagree are researchers at the University of Hawaii who argue that the alleged life-forms are not biological but mineral in nature and "must have formed from a hot, highly pressurized fluid that was squirted into fractures."[22] Dr. William Schopf, a world expert on ancient terrestrial microfossils, also believes that nonbiological processes were involved. He points out that NASA's "Mars microbes" are 100 times smaller than any microbes found on Earth and bear no signs of cells or cavities, which would be crucial indications of life. Like the Hawaii researchers, he thinks the structures are more likely to be minerals.[23] Ralph Harvey of Case

Western University in Cleveland, Ohio, claims that detailed electron microscopic analysis of the alleged microbes "shows a crystal pattern uncharacteristic of life-forms."[24] And researchers at the University of California in Los Angeles have concluded that "the conditions the rock was formed in are not consistent with the theory of life."[25]

In the "pro-life" camp, the work of Professor Colin Pillinger is particularly notable. With his colleagues Dr. Monica Grady and Dr. Ian Wright of London's Natural History Museum he was involved in the discovery of organic material in another Martian meteorite, EETA 79001, and published papers about it in the scientific journal *Nature* before NASA's announcement of possible microfossils in ALH84001.[26] The British researchers initially stopped short of saying that they had found evidence of life. But then in October 1996, they reported that the organic material in the meteorite "contains 4 percent more carbon-12 relative to carbon-13 than exists in neighboring samples of carbonate material. This suggests that the carbon was formed from methane produced by microbial activity." Similar tests on ALH84001 (a fragment of which had been provided by NASA to Pillinger and his colleagues) produced the same carbon isotope ratios.[27]

Of particular interest was evidence that the carbonates in EETA 79001 were far younger than those in ALH84001—not billions of years old but perhaps just 600,000 years old.[28] "Geologically speaking," as one scientist has pointed out, "this is sufficiently recent for there to be a good chance that life may still exist in protected areas on our planetary neighbor."[29]

NASA's Johnson Space Center continues to maintain that the evidence from the Martian meteorites could be "arguably the biggest discovery in the history of science."[30] In London the *Times* predicted that the discovery was the first step in a process "that will profoundly alter our perceptions of the universe and our place in it."[31] In the United States, John Gibbons, the White House Science adviser, commented, "Our notion that life is rare may be revised. Life may be pervasive in the universe."[32] NASA chief administrator Daniel Goldin agrees, stating: "We are on the doorstep to the heavens. We are now on the threshold of establishing, Is life unique to Earth?"[33] The same thought was also clearly in the mind of President Bill Clinton. On the day that the discovery was announced he addressed the nation on television, observing in lyrical tones that confirmation of NASA's findings, if and when it comes,

> will surely be one of the most stunning insights into our world that science has ever uncovered. Its implications are as far-reaching and as awe-inspiring as can be imagined. . . . As it promises answers to some of our oldest questions, it poses others even more fundamental.[34]

We can easily understand why populist politicians might wish to identify themselves with the quest for life on Mars. As Colin Pillinger sums up: "This is what people care about. When I talk to them they only ever want to know if there was life on Mars."[35]

HIDDEN AGENDA?

"NASA has made a startling discovery that points to the possibility that a primitive form of microscopic life may have existed on Mars more than three billion years ago."[36]

With these carefully chosen words, amid much fanfare, news of what had been found in meteorite ALH84001 was first released to the public at a press conference held on 7 August 1996 at the Johnson Space Center in Houston. The speaker was Daniel Goldin, the powerful boss of NASA, who came to the job after spending twenty-five years at TRW, a top-secret defense contractor.[37]

Lobbyists campaigning for more open and accountable government in the United States regard Goldin's presence at NASA as ominous. The appointment was originally made by President George Bush, himself a former director of the Central Intelligence Agency. According to lobbyist and researcher Dan Ecker: "Ever since Goldin has been in charge many of the civilians in NASA have been replaced by former DOD [Department of Defense] people and NASA has steadily been going covert. . . . They have been doing many more Department of Defense missions . . . and remember, Dan Goldin . . . is the only person in charge of a federal agency that I'm aware of that was not replaced under the Clinton administration. That should speak volumes."[38]

Like Ecker, many Americans are convinced that NASA has a hidden agenda and that its policies, and the information it chooses to release to the public, are influenced by factors other than the furtherance of pure science. As we shall see in later chapters, this suspicion has been particularly intense over the issue of the so-called monuments of Mars—notably the pyramids and Face of the Cydonia region. There is a public perception that NASA is involved in a sinister conspiracy to cover up significant evidence about the true nature of these anomalous structures. It has even been suggested that the whole "Mars microbe" extravaganza could have been designed to distract attention from another, more covert Mars story—perhaps to do with Cydonia.[39]

Such speculation sounds like paranoid fantasy. And yet other conspiracies have also been alleged, this time involving the microbes themselves. These allegations stem from reputable scientists working within NASA and cannot easily be dismissed.

MOTIVES

Meteorite ALH84001 is made of rock that has been reliably dated at more than 4.5 billion years old.[40] The life traces identified within it are thought to be 3.6 billion years old. There is good evidence to suggest that the rock was splashed off the surface of Mars 15 million years ago as a result of a collision with a comet or asteroid.[41] It then traveled through space as a piece of cosmic jetsam for millions of years before finally crossing Earth's path just 13,000 years ago and landing amid the ice sheets of Antarctica.[42]

The meteorite's modern history began on 27 December 1984 when it was found in the Allen Hills region of Antarctica. Dark green in color, with tiny rust-red patches in its crevices, it was collected by Roberta Score of the National Science Foundation who recognized it as a meteorite and shipped it to the Johnson Space Center. There, so the official story goes, it was ignored for more than eight years until researchers discovered that it had the classic chemical signature of the SNC class of meteorites and therefore must have originated on Mars.[43]

From 1993 until 1996, sharing almost no information at all with their peers,[44] a group of NASA scientists undertook an intensive investigation of the meteorite. The team was led by David McKay and Everett Gibson of the Johnson Space Center, who later recruited two specialists, Kathie L. Thomas-Keperta of the defense contractor Lockheed Martin and Professor Richard N. Zare of Stanford University, to analyze the meteorite's organic components with a laser mass spectrometer.[45]

According to Dwayne Day of the Space Policy Institute at George Washington University: "As the team became aware of the implications of their research they stopped talking to outside colleagues about it. They were wary of making any comments before they were completely sure of their evidence."[46]

Rather less commendable motives have been suggested by David Des Marais, a scientist at NASA's Ames Research Center. He thinks that the secrecy and exclusive behavior of his colleagues at the JSC probably had more to do with interdepartmental rivalry for funds than with any sense of responsibility or prudence: "There's certainly a lot of competitiveness between NASA centers at the moment with government cutbacks, and so I can imagine why they would want to keep the discovery and announcement all to themselves and have their research and their center making the headlines."[47]

NASA distributes its tasks among many centers. The speciality of Ames, where Marais works, is biological research—notably the chemical and biological experiments carried on the space shuttle. By March 1997, more than seven months after the initial sensational announcements about the Martian microbes, Ames scientists had still not succeeded in persuading the JSC to release a sample from the meteorite for them to study. "We really want to do a chemical analysis on a sample to check for signs of life," Marais commented, "because just about everybody who has looked at the rock up until now has concentrated on its geology. Nobody has investigated its organic chemistry in depth, and we are the best resource to do that."[48]

CREDIT WHERE CREDIT IS DUE

Marais is not the only NASA scientist to have been bypassed by the JSC. Others include Dr. Vincent DiPietro of the Goddard Space Flight Center in Maryland, and Dr. John Brandenburg, who works for the NASA contractor Physical Sciences, Inc.

As we saw in chapter 1, DiPietro is the co-discoverer (with Gregory Molenaar) of the D&M Pyramid in the Cydonia region of Mars. DiPietro's support for the notion that the monuments of Cydonia could be artificial structures—rather than tricks of light and shadow—has for a long while marked him as a rebel within NASA. The same is true of Dr.

John Brandenburg, with whom DiPietro has authored a number of controversial papers about Cydonia.

DiPietro points out that the story of the hunt for life in meteorites from Mars did not begin with the relatively recent efforts of the Johnson Space Center team—who have indeed grabbed all the credit—but with work started as far back as 1966 by the Dutch scientist Bartholemew Nagy. In 1975 Dr. Nagy published a paper on the presence of curious organic compounds in "carbonacious meteorites"—subsequently confirmed to be meteorites from Mars.[49] Fourteen years later, Nagy's findings were corroborated by Colin Pillinger and his team in England in their paper "Organic Materials in a Martian Meteorite," published in *Nature* in July 1989.[50]

Organic materials can be generated by purely chemical as well as by biological processes. In an attempt to establish which process had been involved on Mars, John Brandenburg and Vincent DiPietro undertook a detailed review of the findings of Nagy and Pillinger. By 1994 they had begun to suspect that they had found signs of life. In their paper on the subject, published in May 1996, three months before the Johnson Space Center team went public with their "discovery," they noted that meteorites from Mars are remarkable in that they contain organic material in greater abundance than any other meteoric type. This, they concluded, "could mean evidence for primordial organo-synthesis on Mars and perhaps even primitive biology."[51]

It is odd, and more than just bad manners, that NASA neglected to mention the work of Brandenburg and DiPietro, or the earlier work of Nagy, Pillinger, and Wright, when it made its sensational August 1996 announcement about the discovery of microfossils in meteorite ALH84001. Furthermore, Brandenburg and DiPietro claim that more than a year before the announcement they had personally informed NASA boss Dan Goldin of their own discovery of microfossils in meteorites from Mars. According to DiPietro, they got Goldin's attention for "a couple of minutes" during a conference at the National Academy of Sciences in Washington and put into his hands a dossier of

> writings about the meteorites from Mars which contained organic carbon and fossils. . . . On the very front cover . . . were the pictures of the fossils that were found. He looked at it with some kind of skepticism but also with curiosity. Prior to my putting this into his hands, I had

> addressed this to him in a question so it's in the physical, audiotaped version of that meeting. I had asked him the question about the meteorites, and the fossils that were found within them, and what were NASA's plans for them.[52]

Why therefore did Goldin not acknowledge Brandenburg and DiPietro's findings when he so publicly acclaimed the parallel work of the JSC team?

Brandenburg admits, "Everyone knows we push Cydonia" as evidence of a former civilization on Mars.[53] Since this view has for a long while been unpopular within NASA it has been suggested that Goldin would have been unlikely to have welcomed the prospect of Brandenburg and DiPietro being first past the post with the headline-grabbing proof that life—albeit primitive life—did indeed once exist on the Red Planet.[54]

We are not surprised that Goldin, and perhaps other senior officials at NASA, were well informed about the fossil evidence in Martian meteorites long before that evidence was officially made public. Many large organizations behave secretively as a matter of habit. At the end of August 1996, however, a curious and perhaps significant sidelight was cast on the story by Sherry Rowlands, a thirty-seven-year-old prostitute who claimed to have had an affair with President Clinton's adviser Dick Morris. In press interviews she kept on insisting that Morris had told her about "the discovery of the evidence of a life-form on Mars when it was still a military secret."[55]

LITTLE GREEN MEN

However faint the traces, the smell of intrigue and power politics does hang over the mystery of life on Mars. And yet, what could anybody possibly have to hide?

At the August 1997 press conference Daniel Goldin praised the JSC team for "their dedication, knowledge, and painstaking research," and for making discoveries "that may well go down in history for American science, for the American people and indeed for humanity."[56] At the end of this eulogy he was at pains to emphasize that "we are not talking about 'little green men.' The [fossils] are extremely small, single-cell structures that somewhat resemble bacteria on Earth. There is no evidence or suggestion that any higher life-form ever existed on Mars."[57]

The best that can be said about Goldin is that he seems to have been "economical with the truth" when he gave all the credit for the meteorite discoveries to the JSC team. Could he also have been holding something important back in the second part of his statement when he dismissed the possibility of higher life-forms on Mars? Soon after the press conference, Professor Stanley McDaniel of Sonoma State University made a telling observation about Goldin's presentation: "It's very interesting that as long as it's microbial life, little microbes that are certainly inferior to humans, there's no problem in acknowledging that they may exist, but if it were big or little green men then there's a problem."[58]

There must be a reason for this problem.

The Mother of Life

SCIENCE has yet to explain how, why, when, and where life first emerged. Did it begin on Earth? That is just an opinion. Did it come about as a result of chance combinations of molecules in the "primeval soup"? That is also just an opinion—and so is the opposite opinion that it was the work of a creator. The unvarnished truth, as the biologists Stanley Miller and Leslie Orgel have admitted, is that "we do not know how life began."[1]

Even so, there is agreement on a number of fundamental points. The most important is that "the detection of water in liquid form is the essential indicator for life."[2] According to the biologist Anders Hansson, water, as an inert solvent, "is ideal for biochemical cycling. Szent-Gyorgyi [the Nobel Prize–winning biochemist and discoverer of vitamin C] has called it 'the matrix of life.' Without it life cannot take hold nor the Darwinian evolution begin."[3]

In a realm of science where there are few hard facts this, too, is just another opinion. Nevertheless it is a well-informed opinion and we have no reason to suppose it to be wrong.[4] Until new evidence emerges to the contrary—and because we know that it was so on our own planet Earth—it therefore seems sensible to accept that water is probably a necessary precondition for the emergence of life anywhere in the universe.

Mars today is dead and dry, and cold as hell. With an average overall temperature of minus 23 C, it has no liquid water but only frozen water in the form of ice. Indeed, water in liquid form *cannot* survive on the surface for more than a few seconds in such a climate. It has therefore been baffling to discover, since the era of spacecraft exploration and close-up photography began, that much of the planet shows unmistakable evidence of former oceans, lakes, and rivers, of plentiful rainfall, and of catastrophic floods on a gigantic scale that once scoured its surface.

ICE, DUNES, AND STORMS

Even under the most favorable viewing conditions telescopic observations of Mars can produce misleading results. As we saw in chapter 1, the optical illusion of so-called irrigation canals led Percival Lowell and others in the late nineteenth century to conclude that "Mars is inhabited by beings of some sort or other."[5] The effect was to raise public expections for more than fifty years. Indeed, as late as the mid-1960s, there were many who still confidently expected that the reality of canals would be confirmed by NASA spacecraft. When it was discovered that no canals existed there was widespread disillusionment and a general loss of interest in Mars and its mysteries.

Although the canals are not real, other Martian phenomena, well documented in telescope observations and confirmed by photometric studies, are harder to dismiss as optical illusions. Among these one of the most intriguing is referred to by astronomers as "the wave of darkening":[6]

> Near the edge of either polar cap, a general darkening of the surface markings appears in early spring as the cap begins to recede. The darkening then moves away from the receding polar cap and sweeps toward and crosses the equator in a distinct band of heightened contrast, finally dissipating in the opposite hemisphere. The waves, one in each hemisphere, travel at an apparent speed of about 35 kilometers per day.[7]

The southern polar cap of Mars, at its maximum extent, reaches as far toward the equator as 50 degrees south. The northern cap extends to latitude 65 degrees north, much farther from the equator. By measuring the "reflection spectra" of the caps scientists have discovered what they consist of. The southern cap, by far the colder of the two, is entirely carbon dioxide ice. The northern cap contains fluctuating quantities of carbon dioxide ice but always maintains a permanent remnant, about 1,000 kilometers across, of pure water ice.[8] This is thought to represent "the largest reservoir of available water on the planet."[9]

Surrounding the polar ice, and disappearing beneath it, are what geologists refer to as "extensive layered deposits."[10] Believed to have been carried here by wind, these are cut through by narrow sinuous valleys and circumscribed by the largest sea of sand dunes, or "erg," in the solar system:[11] 'This erg forms a band of windblown sand entirely around the north polar remnant cap. The dunes in this region are spectacular in their regularity over hundreds of kilometers."[12]

From time to time awe-inspiring storms are whipped up on the surface of Mars. For reasons that are not yet understood such storms are usually preceded by a period of sudden local turbulence at certain preferred locations in the southern hemisphere during which tremendous quantities of surface dust get thrown as high as 10 kilometers into the atmosphere. Powerful winds then carry the dust to all parts of the planet, rapidly obscuring its entire surface. Thereafter the intensity of the storm begins to lessen and within a few weeks the atmosphere returns to normal.[13]

EXTRAORDINARY SURFACE FEATURES

Where Earth is mellow and adorned with gentle curves, Mars is a planet of jagged extremes. Its valleys are the lowest in the solar system, its canyons the deepest, its volcanoes the highest. In the absence of an existing sea level, scientists refer to altitudes and depths on Mars in terms of an arbitrary "datum" level. The summit of the giant volcano Olympus Mons, at 27 kilometers above datum, is the highest point on the planet, and the floor of the canyon system known as the Valles Marineris, at seven kilometers below datum, is the lowest point.[14]

Olympus Mons looks like a vision from some dark fairy tale. It is classified by geologists as a "shield volcano" and consists of a circular scab of lava, 700 kilometers in diameter, rising toward a summit caldera 80 kilometers in diameter.[15] The outer edge of the lava scab, around a circumference of almost 5,000 kilometers, is defined by cliffs that drop sheer to the surrounding plains six kilometers below.[16]

Southwest of Olympus Mons is the Elysium Bulge, an immense area of high ground that is surmounted by three volcanoes. The highest of these, Elysium Mons, rises 9 kilometers above the surrounding plains.[17] Southeast of Olympus Mons, at a distance of 1,600 kilometers, begins an even larger upswelling of land. Known as the Tharsis Bulge, it rises 10 kilometers above datum and measures more than 4,000 kilometers from north to south and 3,000 kilometers from east to west—about the size of Africa south of the Congo River.[18] It is in its turn surmounted by three gigantic shield volcanoes—Arsia Mons, Pavonis Mons, and Ascraeus Mons—which are known collectively as the Tharsis Montes.[19] Riding on the broad shoulders of the Tharsis Bulge their peaks rise to 20 kilometers above datum and always remain visible to spacecraft during even the greatest Martian dust storms.[20]

At the eastern edge of the Tharsis Bulge, Mars seems to have been split open by some catastrophic force. Amid a bizarre series of interconnecting box canyons and depressions known as the Noctis Labyrinthis, a tremendous meandering furrow opens in the surface of the planet and runs east—roughly parallel to the equator but between 5 and 20 degrees south of it—for a distance of 4,500 kilometers.[21]

This is the Valles Marineris. Named after *Mariner 9,* the first spacecraft to photograph it, it is up to seven kilometers deep with a maximum width of more than 200 kilometers.[22] By comparison it is four times deeper, six times wider, and more than ten times longer than the Grand Canyon.[23]

At its eastern end Marineris curves northward toward the equator and debouches into a morass of so-called chaotic terrain—a tortured and overturned landscape of blocky remnants, valleys, and fractures that seems like one of the lower circles of Dante's Inferno. From the northern edge of this chaotic zone emerge the deeply etched channels of Simud Vallis, Tiu Vallis, and Ares Vallis (it was in Ares Vallis that NASA's lander *Global Surveyor* touched down on 4 July 1997). All of these channels are very wide and long. They run across the floor of a huge basin known as the Chryse Planitia, where they are joined by other channels, notably Kasei Vallis, which runs out of the north of the central section of the Marineris canyons and is 3,000 kilometers long.[24]

What is striking about the channels, geologists unanimously agree, is that they could only have been caused by floods involving prodigious quantities of water. These floods flowed from the southern hemisphere of Mars into the northern hemisphere at a very rapid rate *because they were draining downhill.*

A DIVIDED PLANET

One of the great mysteries of Mars is that it has two quite distinct and clearly defined areas of relief—the heavily cratered southern uplands, most of which stand at two kilometers or more above datum, and the relatively smooth and uncratered northern lowlands, most of which lie at least one kilometer below datum.[25] The highland and the lowland occupy approximately a hemisphere each, but these only roughly coincide with the present northern and southern hemispheres of Mars. As geologist Peter Cattermole explains:

> The "line of dichotomy" separating these two elevation zones describes a great circle inclined at approximately 35 degrees to the Martian equator.[26]

The main exceptions to the subdatum topography in the "low" northern hemisphere are the Elysium Bulge, entirely inside the northern hemisphere, and a large part of the Tharsis Bulge, which straddles the line of dichotomy.[27] The main exceptions to the above-datum topography in the "high" hemisphere are parts of the Valles Marineris and two stupendous craters, Argyre and Hellas, caused by impacts with comets or asteroids. Argyre is three kilometers deep with a diameter of 630 kilometers. Hellas is five kilometers deep with a diameter of nearly 2,000 kilometers.[28]

These craters together with a third, Isidis, are the largest on Mars. But the planet also has legions of other craters with diameters of 30 kilometers or more, many of which, including one at the south pole, are real behemoths that exceed 200 kilometers in diameter.[29]

All in all, among tens of thousands of smaller craters down to one kilometer in diameter, a grand total of 3,305 craters wider than 30 kilometers have been counted on Mars. Of these, it is difficult to explain why 3,068, or 93 percent, lie south of the line of dichotomy; only 237 such large craters are found north of the line of dichotomy.[30] Equally curious is the fact that the uncratered hemisphere is so much lower in altitude—by several kilometers—than its cratered counterpart.

The reason for this lowland-highland dichotomy, as the geologist Ronald Greely observes, "remains one of the major unsolved problems of Mars."[31] All that is certain is that at some point in its history the planet was afflicted by a cataclysm of almost unimaginable proportions. In chapter 4 we will investigate the causes and consequences of this cataclysm—which a number of scientists suspect may also have been responsible for stripping Mars of its formerly congenial atmosphere and its once abundant resources of liquid water.[32]

WATER, WATER EVERYWHERE

Many of the largest and most damaging Martian craters in the range of 30 kilometers and upward show unmistakable signs of having been made when the planet had a wet and warm environment. Hellas, Isidis, and Argyre in particular have low, indistinct rims and flat floors that several authorities take as evidence of formation when Mars still had a dense atmosphere, rapid erosion, and a stronger magnetic field than it does today.[33] In the same way, acted upon by erosion, craters of great size on Earth "can blend into the landscape in a period of a few hundred years to such an extent as to be practically unrecognizable from the surrounding landscape."[34]

Other large Martian craters, typically measuring 30 to 45 kilometers in diameter, have central peaks, somewhat like gigantic stalagmites, with pits on the summits. Ronald Greely believes that the best explanation for these is that they are "splash" craters and that "water or the atmosphere of Mars, or both, may have been responsible for the form of ejecta."[35]

Planetary scientists Jay Melosh and Ann Vickery have calculated that Mars "probably had an original atmosphere with about the same surface pressure as that of the earth today, and a correspondingly higher surface temperature above the melting point of ice."[36] Their research suggests that the atmosphere was torn away by repeated asteroid impacts: "Because the gravity of Mars is so weak, it is easy for the expanding cloud of vapor from a major impact to blast all of the atmosphere in its vicinity out into space."[37]

In a graphic demonstration of warmer, wetter times, one of the Mars meteorites studied by NASA actually proved to contain a few milligrams of liquid water—the droplet is now kept on display in a sealed glass vial.[38] Moreover, it has been calculated that frozen "subsurface water to a depth of 200 meters may exist on Mars at present."[39] There are even hints that at sufficient depths, close to the planet's inner layers of molten magma, there may be underground hot springs.[40] Theoretically these could vent superheated steam to the surface, and in August 1980, Dr. Leonard Martin of the Lowell Observatory in Arizona reported that two successive images taken by NASA's *Viking* orbiter of an area just south of the Valles Marineris did "suggest an explosive water spout or steam vent."[41] Vincent DiPietro and Gregory Molenaar carried out computer enhancement of these images. They concluded: "Not only did we confirm Dr. Martin's discovery, but we also found a circular compression ring around the center column. . . . The size difference between the images of the two frames indicates the cloud to be rising at a velocity of over 200 feet per second."[42]

The "waterspout" is a controversial matter. But the evidence that Mars possessed vast resources of flowing water in the past is not disputed by scientists and can be seen in plain view in tens of thousands of NASA images. Recently this evidence was subjected to an intensive evaluation by a team in the Exobiology Program Office at NASA. The team included Dr. David Des Marais of NASA's Ames Research Center, Dr. Michael Carr of the U.S. Geological Survey, Dr. Michael A. Meyer of NASA HQ, and the late Dr. Carl Sagan.[43] Their conclusions, which represent the consensus of scientific opinion on this subject, are quoted here at length:

> One of the most puzzling aspects of Martian geology is the role that water has played in the evolution of the planet. Although liquid water is unstable at the surface under present conditions, we see abundant evidence of water erosion. The most intriguing features are large dry valleys, interpreted as having been formed by large floods. Many of the valleys start in areas of what has been termed chaotic terrain in which the ground has seemingly collapsed to form a surface of jostled and tilted blocks 1–2 kilometers below the surrounding terrain. . . . [In Chryse Planitia the] valleys emerge from the chaotic terrain and extend northward down the regional slope for several hundred kilometers. Several large channels to the north and east of [the Valles Marineris] converge on the Chryse basin and then continue farther north, where they merge into the low-lying northern plains. The valleys emerge full size and have few if any tributaries. They have streamlined walls, scoured floors and commonly contain teardrop-shaped islands. All these characteristics suggest that they are the result of large floods. . . . Although most of the floods are around the Chryse basin, they are found elsewhere . . . near Elysium and Hellas. Others occur in Memnonia and western Amazonis. . . .

> Other fluvial features appear to be the result of slow erosion of running water. Branching valley networks are found throughout the heavily cratered terrain. . . . They resemble terrestrial river valleys in that they have tributaries and increase in size downstream. . . . The most plausible explanation for the valleys is that they formed by erosion of running water.[44]

THE SUDDEN END OF A LUSH ENVIRONMENT

Although expressed in the dry language of science, the NASA report nevertheless concerns itself with matters of great significance. It confirms not only that Mars might once have had a wet and relatively warm environment—perhaps even an environment suitable for higher life-forms—but also that this environment seems to have been suddenly swept away.

Other studies have reinforced the same general picture. The major channel system in Chryse Planitia is up to 25 kilometers wide and more than 2,000 kilometers long.[45] It was made by a sudden catastrophic flood that not only shaped its sheer walls but also gouged "cavernous potholes several hundred meters deep" and carved streamlined "teardrop" islands measuring 100 kilometers from end to end.[46] The flood was traveling extremely fast,

> so rapidly as to provide peak discharges of millions of cubic meters per second. Even the dense atmosphere of Earth cannot provide water fast enough to yield such discharges from comparable-sized catchment areas. . . . Only dam bursts have yielded flows of significant macro erosion."[47]

The *volume* of water required to cut the channels has also been estimated. It was very large; Peter Cattermole calculates that it was equivalent to a global ocean more than 50 meters deep.[48] Michael Carr of the U.S. Geological Survey believes that it was equivalent to an ocean 500 meters deep.[49]

Another major flood took place in Ares Vallis. Photographs sent back by NASA's *Pathfinder* lander module in July 1997 show that this immense channel was once filled with "thousands of feet of churning water."[50] According to *Pathfinder* scientist Dr. Michael Malin:

> This was huge. The comparable flood on Earth would be the flood that filled the Mediterranean basin.[51]

Layered deposits of stratified sedimentary material of the kind laid down by the largest terrestrial lakes have been identified in many different locations on Mars. In some places these deposits are five kilometers thick—confirming not only the former existence on Mars of a dense and warm atmosphere in which water could survive in a liquid state but also that the planet's water must have been present for an extremely long period during which Earth-like sedimentation processes occurred.[52] These deductions are strengthened by the compelling evidence, touched on in the NASA report, that rivers flowed in certain regions of the planet for hundreds of millions of years.[53] Moreover, "the existence of run-off channels makes it likely that at one time there was even rainfall on Mars."[54]

THE SHORELINES OF CYDONIA

It is generally believed that these warm and wet conditions last prevailed billions of years ago. However, Harold Masursky of the U.S. Geological Survey has shown that there may have been liquid water on Mars "as recently as a few million years ago."[55] In the U.K., Colin Pillinger and his team have gone further. Their study of Martian meteorites demonstrates that liquid water and primitive life could have existed on the Red Planet just 600,000 years ago.[56] Other researchers, whose work we will consider in chapter 4, are prepared to consider a time frame that is even more recent, with a great cataclysm striking Mars and stripping it of its atmosphere and water less than 17,000 years ago.

Specialists increasingly accept that as well as extensive lakes, "deltas and seas may once have existed on Mars."[57] David Scott of the U.S. Geological Survey has examined "meandering channels, spillways and outlets, spits, terraces, deposits and shorelines" in a number of basins in Elysium, Amazonis, Utopia, Isidis, and Chryse, which he attributes to the presence of former lakes and seas. The Elysium basin, he believes, was once filled with water to a depth of 1,500 meters.[58] Likewise Vic Baker and scientists at the University of Arizona suggest that a great ocean once covered much of the northern hemisphere[59] and support their theory with evidence of ancient shorelines in the low-lying northern plains.[60]

Such features have been identified at latitude 41 degrees north, longitude 9 degrees west,[61] close by the so-called pyramids and Face of Mars in the Cydonia region. According to environmental geologist James L. Erjavec, this region, which lies to the northeast of Chryse Planitia, contains

> areas that look like they're shoreline features, areas where there's erosion, where landslides would occur at the edge of a shoreline, where there may be some erosion of material down below the base of the cliff and sediment has poured into it. Certain erosion features surely indicate that water may have been here in a sizeable quantity. As to what time in Martian history, that still remains to be seen.[62]

The surface of Mars is a palimpsest inscribed with layer upon layer of mysteries. Amid these layers is written the story of the death of a world. It may not have been billions of years in the past, and the fate that afflicted Mars may not have entirely bypassed Earth.

4

The Janus Planet

MARS is a planet of many mysteries, its history only guessed at, its true significance in the solar system as yet unknown. All that is certain is that it was once vibrant with rain and rivers, lakes and oceans, and that it is now barren and dead.

It is the scientific consensus that Mars was killed—*executed* would not be too strong a word—by a stupendous bombardment of asteroids or comets. Thousands of huge craters pockmarking its tortured surface are the silent witnesses to this. And it is thought likely that the same bombardment also caused the cataclysmic floods (described in chapter 3) and then stripped away the planet's formerly dense atmosphere so that liquid water could no longer survive anywhere upon it.[1]

What kind of event could this have been? And what does it say about the nature of the universe in which we live—perhaps even about the predicament of Earth itself—that Mars was so completely rubbed out when it was in its prime?

CLUES FROM THE BODY

We are looking at a murder victim. All we have are photographs and measurements of the corpse and the results of certain scientific tests that have been done on it. These tell us a number of curious things about Mars.

Item 1: Its orbit is highly eccentric and elliptical, following a course that brings it close to the Sun and then very far away from it every year.[2]

Item 2: Its rate of rotation is much slower than it should be.

Item 3: It has almost no magnetic field.

Item 4: Over long periods of time its north-south spin axis seesaws wildly in space, radically changing the angle at which the planet is oriented toward the Sun.

Item 5: There is evidence that the Martian crust may have slipped in one piece around the inner layers of the planet on several occasions in the past—causing landmasses at the poles to be shifted into equatorial zones and vice versa.

Item 6: The vast majority of Martian impact craters, far more than should be the case statistically, are clustered in the hemisphere south of the so-called line of dichotomy (discussed in chapter 3).

Item 7: The northern hemisphere shows only light crater damage and is a vast basin, three kilometers lower in altitude than the south.

Item 8: The line of dichotomy between north and south is physically marked on the surface of Mars by the sheer edge of the upland escarpment. This unique feature runs all the way around the planet in a great ragged circle that crosses the equator at an angle of about 35 degrees.

Item 9: Also unique to Mars is the tremendous chasm of the Valles Marineris—seven kilometers deep, 4,000 kilometers long—that has been torn in its surface.

Item 10: Last but not least there are Hellas, Isidis, and Argyre, the deepest and widest craters in the solar system, weirdly compensated on the other side of Mars by the Elysium Bulge and by the immense Tharsis Bulge—from the eastern edge of which the Valles Marineris bursts forth.

IMPACTS

Let us start with the mystery of the dichotomy. Geologists admit that "despite an ever-increasing awareness of its importance, manifested in intensive research into its nature, mode, and age of formation, there is still no firmly held hypothesis to account for it."[3]

A few scientists favor purely internal, geological process,[4] but the majority agree with William K. Hartmann, writing in the *Scientific American* in January 1977, who pointed out that "an asteroid 1,000 kilometers across striking a primordial planet could have given rise to a fundamental asymmetry in the planet, perhaps by knocking the crust off one side. . . . [This] kind of collision might be involved in the asymmetry of Mars where one hemisphere has many ancient craters and the other has been almost entirely modified by volcanism."[5]

Since the Martian hemisphere lying north of the line of dichotomy has a lower altitude than the southern hemisphere, the automatic assumption has been that the northern hemisphere was struck and lost the outer layer of its crust. The only serious dispute is whether the dichotomy was produced by multiple large impacts in the north[6] or by a "single mega impact,"[7] although both theories present an essentially similar picture of collisions big enough to excavate a basin across an entire Martian hemisphere. Both also assume that there was a time when the north of Mars had a roughly equal number of craters to the south. Then it is supposed that a freak additional bombardment by asteroids (or by one mega asteroid) occurred, for some reason falling only on the north, breaking through its crust, lowering its altitude, and obliterating its preexisting craters. Next, fresh lava welled up from the planet's interior and poured out over the flayed northern hemisphere, covering its wounds and effectively resurfacing it. Subsequently, although occasional asteroids have continued to strike, collisions have become much less frequent and neither hemisphere has experienced any further episodes of intense bombardment.

One important question is sidestepped by both the impact theories: What has happened to the immense volumes of crust, three kilometers deep, that appear to have been scalped from the northern hemisphere? Scientists have calculated that this crustal material would be too massive simply to have eroded away, even over billions of years. As Michael Carr of the U.S. Geological Survey has observed:

> The precise mechanism whereby the former ancient crust has been so extensively destroyed in the northern hemisphere is poorly understood. . . . Erosion alone cannot explain [its] disappearance . . . for there is no sink of sufficient size to accommodate the debris.[8]

The impact theories are also weakened because they call for a freak additional bombardment in the north but are unable to describe any mechanism that would persuasively account for such a bombardment. The best suggestion is that the material flung at Mars was drawn across its orbit because of "perturbations and collisions of bodies in the asteroid belt" probably caused by the atmospheric attraction of Jupiter.[9] But critics say that such perturbations and collisions could not possibly have ejected enough material from the asteroid belt to inflict all the damage that is visible on Mars. Nor is it clear why the damage should have been focused on one hemisphere—the north—with such ferocity that its crust could have been stripped away to a depth of three kilometers. As critics have pointed out:

> *Any* attempt to explain the dichotomy via impact depends on a statistical clustering of impacts in the northern lowland. . . . Unless impacts are significantly more numerous within the lowland than elsewhere, there simply is no reason to expect that the lowland will differ in any way from the rest of the planet.[10]

So could Mars have been hit by "significantly more numerous" impacts in the north than in the south?

There are some who suggest that everything might have been the other way around.

ASTRA

It is the consensus of astronomers that collisions between asteroids and the planets were frequent in the early history of the solar system and have been steadily declining ever since at a uniform and predictable rate. "On any given planet," as a result, it is assumed that "the relative ages are clear, since heavily cratered areas are older than sparsely cratered ones."[11] It is for this reason that the heavily cratered southern highlands of Mars are always referred to as "older" than the "recently resurfaced" plains of the north.[12]

Geographer Donald W. Patten and engineer Samuel L. Windsor have other ideas. They argue that it was not the northern hemisphere of Mars that was the victim of a "freak additional bombardment" (as all other scholars have suggested) but the *southern hemisphere.* They say that this additional shower of cosmic debris is the only reason why the southern hemisphere is more heavily cratered than the north; that is, its surface is *not* older than the northern plains. And, though they do not make the connection themselves, their findings raise an intriguing possibility: *The loss of the northern crust might not have resulted from direct impacts anywhere in the north but could instead have been a "domino" effect from devastating impacts in the south.*[13]

At present there are nine planets in the solar system: Mercury, Venus, Earth, Mars, Jupiter, Saturn, Uranus, Neptune, and Pluto. Patten and Windsor's theory is that there was once also a small tenth planet, orbiting between Mars and Jupiter—in the area where the asteroid belt is now found—and that it came into a collision course with Mars. They name this hypothetical planet Astra and believe that it was drawn toward Mars like a moth to a flame and then destroyed as it entered the larger planet's "Roche limit," a technical term used by astronomers for

> the zone which surrounds any large object of appreciable mass producing a gravitational field at a distance of 2 to 3 radii of the object concerned. In effect it is a danger zone, and any object with a smaller mass or weaker gravitational field entering it will be either swiftly expelled from it electromagnetically, or, more commonly, be subjected to intolerable tidal stress and disintegrated.[14]

The Roche limit is a magical thing, an invisible force field. If its Roche limit is breached, a planet can be expected to defend itself—reaching out, almost like a living being, to destroy the intruder. When this happens the defending planet will suffer serious and perhaps irreversible damage from thousands of fragments of the invader, some very large, that rain down on it. But such damage is likely to be less severe than it would be if there was an actual collision between two intact bodies of planetary size.

Patten and Windsor believe that Astra approached to within 5,000 kilometers of Mars, well inside its Roche limit, and was then torn apart by gravitational and electromagnetic forces—splattering the Martian hemisphere that was facing it with a sudden burst of high-speed projectiles all coming in from the same direction at the same time. The two researchers find plentiful evidence for such an explosion over the southern hemisphere of Mars, pointing out that there is

> an abrupt edge, or rim, for a dramatic drop-off in the density of craters on Mars. That rim [the line of dichotomy] is "where the buckshot ends." It is where the Red Planet's serene [northern] hemisphere begins. This rim is obvious to anyone who is thinking of fragmentation on the Roche limit of Mars. So far, astronomers who fail to think of planetary catastrophism also have failed to see the obvious. The rim rises farthest north on Mars in its northwest quadrant, at latitude 40 degrees north and at longitude 320 degrees west. . . . The southerly extremity of the rim is at latitude 42 degrees south and at longitude 110 degrees west. The rim of craters is not hard to identify if it is expected or anticipated. It is there as it ought to be if Mars experienced a sudden, intense, 15-minute blizzard of fragments bombarding it on one side only.[15]

Very like those who propose selective bombardment of the north, the two researchers' weakest point is that they do not suggest a convincing mechanism that could have put their hypothetical tenth planet, Astra, on a collision course with Mars. Their ideas on this matter rest, in essence, on the belief that the solar system only recently organized itself into its present form and that the orbits of the planets were previously very different.[16]

That few scholars would agree with this aspect of Patten and Windsor's hypothesis does not necessarily mean that they are wrong. Furthermore, even if they are wildly wrong about the mechanism, they still may be one hundred percent right about other things.

They could, for example, be right about the existence of Astra, or something like it. Certainly there is no objection in principle to the notion of an exploded tenth planet as the source of the countless thousands of rocky missiles, some large, some small, that orbit in the asteroid belt between Mars and Jupiter. Indeed, as long ago as 1978 the astronomer Tom Van Flandern of the U.S. Naval Observatory in Washington, D.C., made exactly this case in the planetary science journal *Icarus.*[17] Although admitting that he could think of no means whereby a planet could blow up, he presented persuasive evidence that a tenth planet between Mars and Jupiter could indeed have been destroyed—he thought about five million years ago—and could have been the source not only of the asteroid belt but also of comets entering the inner solar system.[18]

Patten and Windsor's other central idea is of a massive bombardment selectively focused on southern Mars. At the very least, this is no more inherently improbable than the widely accepted notion of a "statistical clustering of impacts" in the northern hemisphere. Moreover a growing body of evidence suggests that the south may indeed have been the target of just such a bombardment.

KILLER PROJECTILES

Hellas, Isidis, and Argyre, the three largest impact craters in the solar system, all lie south of the line of dichotomy.

Centered at 295 W, 40 S, Hellas is an elliptical basin, five kilometers deep, measuring 1,600 kilometers by 2,000 kilometers, so massive that even the ramparts of its rim are 400 kilometers wide.[19] According to Patten and Windsor's calculations, this immense crater formed as a result of an impact with an object measuring 1,000 kilometers in diameter[20]—"as big as Alaska with Washington and half of Oregon thrown in, twice as big as Texas, bigger than most of western Europe."[21]

The Isidis crater measures 1,000 kilometers across and was made, say Patten and Windsor, by an object 600 kilometers wide. Argyre has a diameter of 630 kilometers and was made by an object 360 kilometers wide.[22]

In Patten and Windsor's reconstruction Hellas was the first of the three killer projectiles to reach Mars, screaming down through the atmosphere at a speed of 40,000 kilometers per hour toward a bull's-eye point in the center of the hemisphere south of the line of dichotomy:

> The Hellas fragment hit the crust of Mars a direct blow, from almost vertical. It passed into the internal magma of Mars, creating enormous pressure waves and shear waves. The Hellas fragment did not pass out through the other side of the crust. . . . But the angle of the hit and its velocity did cause sudden, immense internal distress, resulting in a huge pair of bulges in the opposite hemisphere. . . . The Hellas fragment continued to plunge onward, rotating all the way, through the magma of

> Mars. The Tharsis Bulge began to rise, with suddenness, about 100 minutes after Astra fragmented. . . . Simultaneously there were at least two other fragments that penetrated the Martian crust, Isidis and Argyre. In the vicinity opposite the Isidis crater is the second bulge of Mars—the Elysium Bulge.[23]

THE DEATH OF WORLDS

Among tens of thousands of smaller craters, and more than 3,000 craters with a diameter greater than 30 kilometers (including dozens with diameters up to 250 kilometers[24]) Hellas, Isidis, and Argyre are the dark, lurking monsters of Martian topography. Patten and Windsor's estimates of the diameters of the three asteroids that caused these craters—respectively 1,000 kilometers, 600 kilometers, and 360 kilometers—are not correct. We know from studies of impacts on Earth that an object 10 kilometers in diameter can make a crater nearly 200 kilometers wide. More accurate estimates of the Martian impactors suggest diameters in the range of 100 kilometers for Hellas, 50 kilometers for Isidis, and 36 kilometers for Argyre.[25]

For a planet the size of Earth (and Mars is not much more than half the size of Earth), it is important to understand that a collision with any object wider than about one kilometer is a catastrophic event. Indeed, extensive damage has been caused on Earth by much smaller objects. The famous Barringer Crater in Arizona, which is 180 meters deep and just over a kilometer wide, was gouged out by an iron meteorite no larger than 50 meters in diameter.[26] The so-called Tunguska Event of 30 June 1908 was an aerial explosion over Russia of a fragment of a comet or asteroid measuring 70 meters across and traveling at 100,000 kilometers per hour.[27] Estimated to have occurred at an altitude of about six kilometers above the Siberian plains, this vast explosion flattened more than 2,000 square kilometers of forest, completely incinerated a central region of 1,000 square kilometers, and ignited people's clothes as far afield as 500 kilometers from the epicenter.[28] Seismic shocks from the Tunguska Event were measured at a distance of more than 4,000 kilometers, and so much dust was thrown up into the atmosphere, blocking out sunlight, that the earth's surface temperature was measurably reduced for several years afterward.[29]

The Tunguska object was 70 meters across and, mercifully, exploded over an unpopulated area before colliding with Earth. Sixty-five million years ago, another object, this time *10 kilometers* across, crashed into the northern end of the Yucatan Peninsula and the Gulf of Mexico with an explosive force that is estimated to have been a thousand times more powerful than all the nuclear bombs and missiles currently stockpiled on earth. It gouged out a crater 180 kilometers in diameter, sent up a dust cloud that blotted out the sun for five years, and created seismic instabilities that wracked the entire planet for decades with aftershocks and volcanic erruptions.[30]

The notorious "K/T boundary Event" wiped out the dinosaurs, and 75 percent of all other species then living on Earth.[31] It has been aptly described as

> one of the greatest disasters that has ever hit our planet. . . . It was the equivalent of a rock the size of Mount Everest, traveling ten times faster than the fastest bullet, producing an impact so severe that the entire Earth shifted in its orbit by a few dozen meters.[32]

That a "rock the size of Mount Everest," with a diameter of just 10 kilometers, could have caused a planet-wide cataclysm that almost ended life on Earth is surely a chilling thought. Asteroids and comets measuring 10 kilometers or bigger are relatively common in the solar system, and we shall see in part 4 that many of them hurtle along on potentially disastrous Earth-crossing orbits.[33] Astronomers refer to them as "Apollo objects"[34] and believe that some may reach 100 kilometers in diameter.[35] Such giants are thought to be rare, but it is widely understood that a collision with one of them would be a world-killing event in which it is unlikely that any form of life would survive.

It is worth repeating that the object that excavated the Hellas crater on Mars had a diameter of 100 kilometers. The Isidis object had a diameter of 50 kilometers. The Argyre object had a diameter of 36 kilometers.

As each of these huge interplanetary dumdum bullets was large enough to have killed Mars on its own, it is not hard to imagine what the global consequences of three such impacts must have been. Indeed, imagination is superfluous, because we have the NASA photographs of the ruined corpse of Mars to tell us the whole story. At the risk of overextending the metaphor, what these photographs suggest is that the "victim" was first hit from the south at point-blank range with the cosmic equivalent of a blast from a 12-bore shotgun—hence the thousands of craters clustered to the south of the line of dichotomy—and that the "killer" then finished off the job with three single shots from a large-caliber rifle.

ENERGY WAVES

Sixty-five million years ago, at the moment that the 10-kilometer-wide comet or asteroid that destroyed the dinosaurs hit Earth, tremendous shock waves were sent surging around the planet from the point of impact in the Gulf of Mexico. Geologists do not think it is an accident that almost exactly on the opposite side of the globe, at exactly the same time, an extraordinary burst of volcanic activity occurred in India. Wide-scale seepage of molten magma through fissures in the earth rapidly built up a great shield of basaltic lava—nearly a thousand meters high and thousands of square kilometers in area—which cooled to form the Deccan Traps. "Shock waves rippling out from the impact," observe science writers John and Mary Gribbin, "would have tended to focus together again in just about that part of the world."[36]

Patten and Windsor's argument is that much the same thing, only a hundred times worse, happened on Mars—that the Tharsis Bulge swelled up in reaction to the Hellas impact and that the Elysium Bulge was a reaction to the Isidis impact. The shock waves are estimated to have been of such magnitude that they would not merely have passed around the planet but would have punched directly through it, ahead of the penetrating asteroids that cut into Mars like augurs. Indeed, it has been calculated that from their points of entry south of the line of dichotomy the Hellas, Isidis, and Argyre asteroids could possibly have traveled a distance of some 5,000 kilometers before coming to a halt inside the opposite, "serene" hemisphere north of the line of the dichotomy.[37] There they would have released gigantic pressure waves that would have rushed upward to the surface at about 5,000 kilometers per hour.[38]

It is an entirely reasonable proposition, well supported by the Deccan Traps precedent on Earth, that this could have produced sufficient volcanic activity at the surface to account for Tharsis and Elysium—and probably for Olympus Mons as well. In addition, Patten and Windsor suggest that the sudden need for Mars to absorb and "digest" the mass and kinetic energy of the three large asteroids may have brought it close to total destruction. It was not enough for the planet to vent magma at Elysium and Tharsis. The pressure and expansion called for further release, and from the eastern edge of Tharsis the planet burst its seams along fully one-quarter of its circumference, forming the formidable gash that we know as the Valles Marineris.[39] This vertiginous canyon system reaches depths of seven kilometers—too deep, according to authorities like Peter Cattermole, to be explained by internal geological processes.[40]

Is it possible that one other thing—more devastating than all the rest—could have happened to Mars as a result of the three gigantic impacts it suffered? Is it possible that the hammer blows that it received from within, emanating from the south, could have transmitted sufficient energy to the north to shake loose the crust?

This was almost exactly the scenario envisaged by William K. Hartmann in *Scientific American*—that a collision with just one very large impactor could theoretically account for the Martian "asymmetry." As we have seen, it has always been assumed that such a collision—or multiple collisions—would have occurred in the northern hemisphere. But recent research supports the notion that tremendous energy pulses transmitted from south to north during the Hellas, Isidis, and Argyre impacts could have done the job just as effectively. This research has shown that even shock waves from relatively small impacts have caused the surface of Mars "to bounce, flicking boulders up to 15 meters across into space."[41]

Hellas, Isidis, and Argyre were not small impacts. The possibility cannot be ruled out that their combined mass and momentum could have "bounced" the entire northern hemisphere vigorously enough to flick a three-kilometer-thick layer of its crust into space.

DISORDER AND DISTURBANCES

Hellas alone was 100 kilometers in diameter. Combined with the Isidis and Argyre impactors, it is not inconceivable that it may have "carried so much energy and momentum" that on colliding with Mars, "it could have tilted it, speeded up its spin, slowed down its spin, destroyed a satellite, or perhaps even have left rings of material around it after breaking up under gravitational forces."[42]

NASA observations going back as far as *Mariner 4* suggest that the Martian orbit, which the reader will recall is unusually elliptical, "has been seriously disturbed and the planet's structure severely strained at some time in the past."[43] Furthermore, telltale fractures on the Martian crust indicate that there has at some point been a significant change in "the planet's rotational equilibrium figure"—that is, in its rate of spin.[44] The laws of celestial mechanics dictate that it should be revolving once every eight hours; instead a complete revolution takes almost twenty-five hours.[45] Such a change appears to have been far too large to have been caused by tidal interaction with Phobos and Deimos, the two tiny moons of Mars, and scientists recognize that "some other cause" must be sought.[46]

Might that same cause have had something to do with another oddity of Mars?—namely the fact that the tilt, or obliquity, of its spin axis is subject to wild fluctuations. Presently at about 24 degrees, its "normal" range is already very large, varying from 14.9 degrees to 35.5 degrees over cycles of just a few million years.[47] In 1993, Jihad Touma and Jack L. Wisdom of the Massachusetts Institute of Technology discovered that "the tilt can also change abruptly. Excursions of the tilt axis through a range of as much as 60 degrees may recur sporadically every ten million years or so."[48]

Another curious characteristic of Mars is that it has almost no magnetic field, although there is undisputed evidence that it did once possess a strong one.[49] And last but not least, there is evidence of a major, possibly rapid and possibly violent one-piece slippage of the entire Martian crust around the inner layers of the planet. For example, typical mantled and layered polar deposits have been found 180 degrees apart at the equator—that is, in positions antipodal to one another—as would be expected with former poles.[50]

INTERPLANETARY VISITORS

What set the Martian crust in motion and its axis rocking, and snuffed out its magnetic field and drastically slowed down its rate of spin? Was it the same event that brutally cratered the south of the planet and scalped the north down to a depth of three kilometers? And when did these things happen?

Patten and Windsor suggest that many of the answers lie with their hypothetical tenth planet, Astra. Such a body could certainly have disturbed the orbit of Mars—and slowed down its spin—if, as supposed, it had exploded inside the planet's Roche limit. This is by no means an unorthodox position. Hartmann too speaks of the possibility of "a large interplanetary body" entering the solar system[51] and envisages how it might have trespassed the Roche limit of one of the planets and been "torn apart by tidal forces."[52] Where Patten and Windsor do fly in the face of conventional wisdom, however, is in their proposed chronology. They assert that the timing of the Astra cataclysm was "thousands of years ago, not millions."[53] Subsequently they narrow down the window to a period "neither earlier than 15,000 B.C. nor later than 3000 B.C."[54]

In their important study *When the Earth Nearly Died,* D. S. Allen and J. B. Delair also propose a massive interplanetary visitor—to which they give the name Phaeton. Like Patten and Windsor they believe that its appearance was extremely recent and that it passed close to Mars and Earth approximately 11,500 years ago.[55] As to the precise nature of the object, they suggest that "Phaeton was spawned in an astronomically-near supernova explosion," and that "Phaeton was a portion of exploded astral matter."[56]

Other authorities who make a related case include the eminent Oxford University astronomer Victor Clube and his colleague William Napier whose extraordinary work we will examine in part 4. They present evidence that a giant interstellar comet wandered into the solar system and began to fragment less than 20,000 years ago spreading ruin among the planets.[57]

TWO PLUS TWO EQUALS FIVE?

Until rock samples can be returned to Earth for radiometric tests, all proposed chronologies for the planet Mars should be regarded with skepticism. This is because the only dating procedure presently available to researchers is to pore over orbiter photographs and *count the craters* on features for which they wish to establish an age. As the reader will have gathered, the basic assumption behind this sort of abacus-level science is that impacts with asteroids and meteorites have occurred at a predictable rate over the last four billion years or so, with the largest number of impacts being registered early in the history of the solar system.[58] Accordingly, heavily cratered areas are always judged to be older than lightly cratered areas, and because Mars is heavily cratered south of the line of dichotomy it is assumed that most of the cratering there must have occurred billions of years ago.

Yet crater counting has some severe and perhaps fatal flaws. Peter Cattermole points out that it cannot give *absolute* dates, only *relative* dates.[59] This is because it is honestly impossible from photographic evidence alone to assess how long ago an impact occurred. The most that crater counts can do is tell us that "some feature is probably older or younger than another feature, but we cannot say by how much or what the age of each feature is."[60] Because of this grave weakness, the method cannot make any allowance for the possibility, envisaged by Patten and others, of a sudden erratic, unpredictable blizzard of missiles hitting one hemisphere of Mars all at once, creating huge numbers of craters in a very short time, *perhaps recently,* thus giving the illusion of great age to features that are in fact young.[61]

Could it be such an illusion that has convinced most scientists that Mars was last massively bombarded billions of years ago? Could a tremendous mistake have been made?

LOST CIVILIZATIONS

The notion that the terminal Mars cataclysm might have occurred recently—perhaps less than 20,000 years ago—is an astronomical heresy that raises peculiar resonances for us.

In earlier books we have shown that an enormous cataclysm occurred on Earth in precisely this period.[62] It was then that the last Ice Age came abruptly and disastrously to an end. No scientist has ever explained how or why this tremendous change occurred. The only certainty is that the sprawling ice caps of the Wurm and Wisconsin glaciations, which had enshrouded northern Europe and North America for at least 100,000 years, suddenly went into a ferocious meltdown—and that this began around 17,000 years ago. The next eight thousand years witnessed catastrophic floods, earthquakes, volcanic activity, and an overall rise in sea levels of more than 100 meters.[63]

By the time the worst was over the face of the earth had changed almost beyond recognition: former coastlines, islands, and land bridges had been inundated, and many animal species had passed into extinction. Emerging from the mud and the ashes, the survivors included a small ragged remnant of humanity.

Among the most treasured baggage that these surviving humans carried with them were memories—in the form of myths—of far-off times "before the Flood" when a great civilization flourished and the world was ruled by god-kings with mysterious powers and strange technology. In *Fingerprints of the Gods* and in *Message of the Sphinx* we showed that these myths, which are astonishingly consistent from culture to culture, could reflect a profound historical truth. An advanced civilization could indeed have arisen during the last Ice Age—only to be destroyed by the global flood that brought the Ice Age to an end. Some of the oldest myths and scriptures invite us to consider the possibility that the sacred wisdom and technical knowledge of this antediluvian civilization might not have been entirely lost in the cataclysm—that indeed a concerted effort might have been made to ensure that the best parts of an extraordinary legacy would be preserved.

We have traced the theme of hidden knowledge through a labyrinth of ancient sites in widely scattered regions of the world.[64] Our travels have convinced us that among these sites is one that is paramount—Egypt's Giza necropolis, the sacred domain of the three Great Pyramids and the Great Sphinx. We have made the case that elements of this site may be far older than the 4,500 years allocated to them by orthodox scholars, some dating back as far as 12,500 years, and we have shown that the pyramids and the Sphinx are terrestrial models of the constellations of Orion and Leo as they last appeared in the sky above Egypt 12,500 years ago.[65] We have also investigated traditions of a "hall of records" at Giza—perhaps hidden in the bedrock under the Sphinx, perhaps in a concealed chamber in the Great Pyramid—where the ancient Egyptians believed that sacred writings from before the flood were stored.

We are not prepared to rule out the possibility that such a repository—a time capsule from an antediluvian civilization—could still exist and may yet be found.[66] Nor are we prepared to rule out the possibility, suggested by the work of Clube, Napier, Allen, and Delair, that the cataclysm that struck Earth at the end of the last Ice Age could have occurred in the same epoch as the cataclysm that destroyed Mars—*and might have had the same cause.*

We therefore naturally find it curious, and will investigate the matter in later chapters, that the ancient Egyptians envisaged a profound connection between Mars and Earth and, more specifically, between Mars and the Great Sphinx of Giza. The planet and the monument were both seen as manifestations of Horus, the divine son of the star-gods Isis and Osiris. The planet and the monument were also both called by the same name, Horakhti, meaning "Horus in the Horizon." Mars in addition was sometimes known as Horus the Red, and the Great Sphinx, for much of its history, was painted red.[67]

What really died on the Red Planet during its last great cataclysm?

We already know that the solar system lost something infinitely more precious than just a barren and empty world when the murderous barrage of cosmic debris slammed into Mars. We know that until the moment of its execution the planet possessed a strong magnetic field and a dense Earth-like atmosphere that permitted the formation of oceans, lakes, and rivers. We know that there had been frequent heavy rains on Mars and that vast quantities of water are still locked up as ice at its poles and beneath its surface. We know that many tantalizing hints and traces of organic life processes have been encountered.

We also know that there is a gigantic Sphinx "face" on the plains of Cydonia, close to the shores of a former ocean, associated with a group of immense pyramidal structures.

Are these just tricks of light and shadow playing with weird geology?

Or is the most staggering revelation of the millennium about to unfold?

PART TWO

The Mystery of Cydonia

Close Encounter

HUMANITY'S close encounter with Mars and the current search for life there may ultimately be looked back upon as a seminal moment of history. So far as we know, such an encounter has never happened before. Nevertheless, since NASA's physical exploration of Mars is the end-product of more than a century of international endeavor, our reactions to what is found there will inevitably be influenced by entrenched ideas.

Scientific interest in the possibility of life on Mars seems to have begun in 1877 when the Italian astronomer Giovanni Schiaparelli announced a startling new discovery. He had observed a network of crisscrossing single and double lines on the Martian surface—giant grooves, *canali* in Italian, a word that was translated loosely into English as "canals."[1] Schiaparelli's findings were widely hailed at the time as evidence for the existence of an intelligent extraterrestrial civilization on our neighboring planet. Among those who found themselves electrified by the discovery was the American Percival Lowell, a rich Harvard graduate with an interest in astronomy.

Lowell read up on Schiaparelli's canals in *La Planete Mars,* a book by the French astronomer Flammarion,[2] and was inspired to build an observatory to study the planet under clear skies and at high altitude in the Arizona city of Flagstaff.[3] He referred to his work there as a "speculative, highly sensational and idiosyncratic project."[4] Its goal, he said,

> may be put popularly as an investigation into the condition of life on other worlds, including last but not least their habitability by beings like [or] unlike man. This is not the chimerical search some may suppose.

On the contrary, there is strong reason to believe that we are on the eve of a pretty definite discovery in the matter.[5]

CANALS AND FLYING MACHINES

Lowell died in 1916, having arrived at no definite discovery, but his views of the nature of Martian life were to have lasting effects, capturing the public imagination for decades.

One popular theory of Lowell's was that the Martian canals brought water from the frozen polar ice caps to an ancient civilization, far older than any human civilization, in the arid vastness of the planet's tropical and equatorial deserts.[6] He also proposed that certain dark fluctuating patches visible on the surface of Mars were vegetation.

Lowell was using the most up-to-date equipment to make his discoveries, and his announcements caught the mood of the world at the fin de siècle—an openness to new ideas such as occultism and spiritualism, which naturally favored the possibility of life on other planets.[7]

The widespread interest in occultism and extraterrestrial life lay behind the success of the prodigious French writer Camille Flammarion. In 1861, aged nineteen, he wrote a book entitled *La Pluralité des Mondes Habités,* which argued for the probable existence of extraterrestrial life. It became an instant bestseller, as did his later work, *La Planete Mars* (1892), the book that directly inspired Lowell. In it Flammarion states:

> The actual conditions on Mars are such that it would be wrong to deny that it could be inhabited by human species whose intelligence and methods of action could be far superior to our own. Neither can we deny that they could have straightened the original rivers and built up a system of canals with the idea of producing a planet-wide circulation system.[8]

The ideas of Schiaparelli, Flammarion, and Lowell stoked Mars fever in the final years of the nineteenth century. In 1898 H. G. Wells cashed in on this with his tale of the Martian invasion of Victorian Britain, *The War of the Worlds.* Then in 1902 the eminent Swiss psychologist Carl Gustav Jung published his doctoral dissertation *On the Psychology of So-called*

Occult Phenomena. In it he subjected his cousin Hélène Preiswerk—who was in the habit of falling into mediumistic trances—to a detailed psychological analysis.

In her trances Hélène often talked of journeys to Mars:

> Flying machines have long been in existence on Mars. The whole of Mars is covered with canals, the canals are artificial lakes and are used for irrigation. The canals are all flat ditches, the water on them is very shallow. There are no bridges over the canals, but that does not prevent communication because everybody travels by flying machine.[9]

Clearly Flammarion and Lowell's Mars was entering humanity's psyche at a most profound level. Here, a fourteen-year-old, uneducated Swiss girl, in unconscious utterances, was revealing the preoccupations of an era.

In 1902, the same year that Jung's thesis was published, a prize was offered to the first person to make contact with an alien life-form. There was one stipulation: contact with Martians was not included, for the simple reason that this was thought to be too easy. In 1911, nine years after the competition was launched, an article appeared in the *New York Times* headlined "Martians build two immense canals in two years."[10]

EXPERIMENTS

The belief that Mars could be, if not inhabited, then at least habitable was sustained among laymen and scientists alike until the second half of the twentieth century. For example, in the early 1960s the popular British astronomer Patrick Moore and a microbiologist, Dr. Francis Jackson, sought to test the possibility of life on Mars by conducting simple experiments:

> We built a Martian Laboratory, filled it with what we thought to be the correct atmosphere—nitrogen, with a pressure of 85 millibars—and gave it the right temperature range between day and night. When we grew things in it, the results were interesting. A cactus fared badly, and after a single Martian night looked decidedly worse for wear, but more simple organisms did better, and we felt quite encouraged.[11]

Similarly, the late Carl Sagan built what he called "Mars jars" in which these experiments were repeated.[12] His results were similar—some microbes actually grew if a little water was present.

But any optimism over such results was soon to be crushed when space probes in the mid-1960s sent back images of Mars as a barren and frozen lifeless hell.

ROCKETING TECHNOLOGY

In 1926 Robert Hutchings Goddard (NASA's Goddard Space Flight Center is named in his memory) built the forerunner of the space rockets that we are familiar with today—though his small prototype traveled only 60 meters before crashing and could reach a top speed of just 100 kilometers per hour.[13] He was the first person to test and prove the theory that rockets could be used to leave the earth's atmosphere and even travel to other planets—a view first proposed by a Russian schoolteacher named Konstantin Eduardovich Tsiolkovsky in the late nineteenth century, and further refined by the German Hermann Oberth in 1923.

During World War II, the rocket was developed as a weapon by the Nazis. Their V-2 relied upon and improved Goddard's technology. Three years after the end of the war a two-stage V-2/WAC Corporal combination bettered Goddard's distance phenomenally, reaching a height of four kilometers.[14]

THE SPACE RACE

If the Second World War was a catalyst to rocket science, then the Cold War was a thousand times stronger. With the threat of nuclear annihilation hanging in the air, the American rocket program—led initially by Wernher von Braun—waged a guerrilla campaign of intellect and design with its Russian counterpart, headed by Sergei Korolov. On both sides of the Iron Curtain, masses of government funding went into improving the propulsion systems for atomic weapons.[15] On 4 October 1957, an offshoot of all this research and development allowed the Russians to send humanity's first ever satellite, *Sputnik I,* into orbit. The "space race" had begun.

Russia scored the next triumph, too, by putting the first man into space. The successful mission of Yuri Gagarin in *Vostok* completely obscured the efforts of the American space program, which had been hastily kick-started in 1958 in response to the launch of *Sputnik.*

In that year NASA, the National Aeronautics and Space Administration, was founded.[16] The United States also launched its own satellite, *Explorer 1,* sending it into orbit on a Jupiter C rocket provided by the U.S. Army at the Jet Propulsion Laboratory, Pasadena, California. Then came Gagarin's great success in 1961. Soon afterward President John F. Kennedy pledged that NASA would put a man on the moon by the close of the decade.

Kennedy's pledge was fulfilled on 20 July 1969 when Neil Armstrong took "one small step" out of the *Apollo 11* lander onto the surface of the Moon—the thirty-third American probe to be sent there. This "giant leap for mankind" was a leap fueled by international competition and war. It was a leap into a new order of discovery, a leap that would give us a new view—that of Earth hanging in space, beautiful and unified, undivided by political and national boundaries.

THE MARS MISSIONS

It was the Russians who sent the first probe to Mars—the appropriately named *Mars 1,* which was launched on 1 November 1962. It is believed to have approached to within 195,000 kilometers of the planet, but all contact with it was lost on 21 March 1963 before it could send back any observations.[17] Its fate is one that has mysteriously dogged many missions to Mars.

NASA's first Mars probe was *Mariner 3,* which was launched on 5 November 1964. Like its Russian predecessor, it too was a failure, going out of control early in the mission. Apparently its protective fiberglass shroud failed to eject on exiting the earth's atmosphere, thus making it too heavy to stay on its projected course.[18]

AMERICAN SUCCESS

Three weeks and two days later, on 28 November 1964, *Mariner 4* was launched. First blood went to the Americans as the craft sent back twenty-one photographs and vital new information, getting within 10,000 kilometers of Mars.[19] The murky images picked up the planet's densely cratered and lifeless surface. They were man's first glimpse of Mars at close range—a glimpse that shattered many myths.[20]

Just two days after *Mariner 4*'s launch, the Russian *Zond 2* attempted to reverse the disastrous fate of *Mars 1*—and failed. In late spring of 1965 all contact with it was lost.

On 24 February and 27 March 1969, NASA launched two new Mars probes—*Mariner 6* and *7*. *Mariner 6* traveled to within 3,390 kilometers of Mars and took 76 pictures. *Mariner 7* approached to 3,500 kilometers and sent back 126 pictures.[21]

WASTELAND

The early Mars missions were a disappointment to many. Bugged by technical failures, and overshadowed by the high-profile moon missions, the images they returned were not exciting. There was no vegetation—the dark patches of Mars proved to be "albedo areas" in which the red topsoil had blown away to reveal darker rocks underneath. There were no canals. Mars was heavily cratered and apparently very old.

The first successful probe, *Mariner 4,* revealed that the Martian atmosphere was not nitrogen (as Moore and Jackson had proposed) but largely carbon dioxide, as were, in all probability, large areas of the frozen ice caps. Liquid water could not exist on Mars since the surface pressure was much lower than previously thought—lower than 10 millibars, not around 85.[22] It was an inhospitable nightmare world—drab and lifeless, apparently devoid of any interesting features. And theories such as Lowell's were dispelled like phantasms in the cold, hard light of the Martian day.

As a NASA spokesman said:

> "We've got superb pictures. They're better than we could ever have hoped a few years ago—but what do they show us? A dull landscape as dead as a dodo. There's nothing much left to find."[23]

The next decade would prove this view of Mars as wrong as Lowell's had been.

6

A Million to One

The storm burst upon us six years ago now.

As Mars approached opposition, Lavelle of Java set the wires of the astronomical exchange palpitating with the amazing intelligence of a huge outbreak of incandescent gas upon the planet. It had occurred toward midnight of the 12th, and the spectroscope, to which he had at once resorted, indicated a mass of flaming gas, chiefly hydrogen, moving with an enormous velocity toward this earth. This jet of fire had become invisible about a quarter past twelve. He compared it to a colossal puff of flame, suddenly and violently squirted out of the planet, "as flaming gas rushes out of a gun."

A singularly appropriate phrase it proved. Yet the next day there was nothing of this in the papers, except a little note in the *Daily Telegraph*, and the world went in ignorance of one of the gravest dangers that ever threatened the human race. I might not have heard of the eruption at all had I not met Ogilvy, the well-known astronomer, at Ottershaw. He was immensely excited at the news, and in the excess of his feelings invited me up to take a turn with him that night in a scrutiny of the Red Planet. . . .

He was full of speculation . . . about the condition of Mars, and scoffed at the vulgar idea of its having inhabitants who were signaling us. His idea was that meteorites might be falling in a heavy shower upon the planet, or that a huge volcanic explosion was in progress. He pointed out to me how unlikely it was that organic evolution had taken the same direction in the two adjacent planets.

> "The chances against anything man-like on Mars are a million to one," he said.[1]

In early 1998, exactly a century after H. G. Wells wrote these words in the first chapter of *The War of the Worlds,* NASA's Mars *Global Surveyor* probe was scheduled to begin mapping the surface of the Red Planet.

This is not a new task—Mars has been thoroughly mapped before by both American and Russian probes. However, *Global Surveyor* has been designed to send back to Earth the most detailed images of the Martian surface yet taken from space.[2] The possibility cannot be ignored that what it might find could irrevocably change mankind's future and all our notions about the past.

For against all expectations, it would seem that there is something "man-like" on Mars. A century after Ogilvy stated his odds we may be poised on the edge of a discovery beyond Wells's wildest dreams—a discovery worthy of a Schiaparelli or a Lowell that scientists claim is an illusion, but which, if it is not, is a profundity beyond our comprehension. Moreover, to echo Lowell: "There is strong reason to believe that we are on the eve of a pretty definite discovery in the matter."[3]

The man-like something is the Face on Mars—the colossal mound that rises nearly 2,600 feet above the barren Cydonia plain, on the shoreline of a long-vanished Martian sea, a mound seemingly carved into immense humanoid features staring hauntingly up at us.

And yet, like the "flaming gas" of Wells's fictional tale, this mysterious object, and the many others that surround it on the Cydonian and Elysium plains—the implications of which could be, pardon the pun, astronomical—remain relatively unheard of and unstudied. This is because the majority of scientists, like Wells's Ogilvy, remain firm in their beliefs that the chance of there ever having been man-like life on Mars is still "a million to one."

A century on, are our modern-day Ogilvys about to be forced to change their views in the light of new evidence? Will *Mars Global Surveyor* confirm that fact is indeed stranger than fiction? For it is a fact that both the principal Mars probes of the 1970s—*Mariner 9* and *Viking 1*—photographed objects on the surface of the planet that have been claimed as evidence for the existence of intelligent life on another world.

MAY 1971

The 1960s proved a pioneering yet ultimately disappointing time for Mars research, with initial feelings of enthusiasm punctured by the early *Mariner* images of the Red Planet as a dull, lifeless, cratered hell. For some time nobody knew that the pictures taken by those early missions had completely missed the varied, wondrous geological features that make Mars such an amazing and mysterious planet.

The end of the 1960s freed the superpowers from their race to the Moon. They promptly renewed their fervor for Mars, sending a total of five spacecraft within a 22-day period in May 1971.

Two of the craft, *Mariner 8* and *9,* were American. The function of *Mariner 8* was to map Martian topographical features, scanning 70 percent of the planet's surface from a highly inclined orbit. The idea was to photograph Mars with the sun very low on the horizon, throwing long shadows. *Mariner 9,* on the other hand, would position itself for a high sun angle to take pictures of albedo features in the equatorial regions.[4]

Mariner 8 was launched on 8 May 1971. Shortly after takeoff, due to a guidance system malfunction, the second stage of the Atlas-Centaur rocket carrying the probe separated from the primary, but failed to ignite. This probe plunged into the Atlantic Ocean, 360 kilometers north of Puerto Rico.

It was left to *Mariner 9* to make up the loss and its role was adapted to include aspects of its failed counterpart's mission. The new plan was to place the craft in an intermediary orbit, inclined at 65 degrees to the equator, and at the minimum altitude of 1,350 kilometers.

Mariner 9 took off from Cape Kennedy (later Canaveral) 22 days after *Mariner 8*'s demise. It would, however, not be going alone: Just two days after the loss of *Mariner 8* a Soviet Mars orbiter had been launched from Baikonur in Kazakhstan. Like its American counterpart, due to a stupid mistake in the computer systems, it failed to leave the earth's orbit. Before the end of May, however, two more Soviet craft, *Mars 2* and *Mars 3*—each consisting of an orbiter with a detachable lander—had been launched successfully.

So the summer of 1971 saw three interplanetary craft safely leaving Earth's sphere of influence and heading silently toward our red neighbor.

DUST STORM

A few months earlier, in February 1971, an astronomer at the Lowell observatory in Flagstaff, Charles F. Capen, made a prediction concerning the weather on Mars. He thought it probable due to the position of Mars at that time—in "perihelic opposition"—that a dust storm could arise toward the end of the summer. Sure enough, on 21 September, as the three craft were approaching Mars, a small cloud developed over the Hellespontus region.

When *Mariner 9* turned on its TV camera on 10 November (having overtaken its Russian rivals to be within 800,000 kilometers of Mars), it revealed a planet whose surface was completely obscured by a violent global dust storm. Nothing could penetrate the veil of dust. And so *Mariner 9* performed an operation that would secure its place in the immortal heaven of the history of space exploration. It switched off its camera and waited.

The two Soviet craft, *Mars 2* and *3,* were modeled on the *Venera* orbiter-lander craft that the Russians had deployed on the surface of Venus in the 1960s. The *Venera* missions had been moderately successful, sending back information from the landers during descent, but losing communication after they reached the surface. If the lander modules on the Mars probes were equally successful then they would be a sensation and would overshadow anything achieved by *Mariner 9*—a dedicated orbiter with no lander module.

Mars 2's lander failed to make a smooth descent. On 27 November 1971 it crashed into the Martian surface at a point north of Hellas, 44.2 S, 313.2 W.

Five days later the *Mars 3* lander deployed. On the way down it transmitted blank frames for twenty seconds before all contact was lost. Having landed in the midst of a violently destructive dust storm, it is thought that its parachute was dragged by 140 meter-per-second winds and that it was smashed to bits.

MARINER 9

As the Mars landers were consumed in the global dust storm below, *Mariner 9* drifted silently in orbit, dormant, conserving its energy.

Meanwhile the *Mars 2* and *3* orbiter modules, from which the unsuccessful landers had been deployed, snapped away at the Red Planet in a whir of irreversible, preprogrammed activity—and sent back to a devastated Russian team picture upon picture of dust clouds.

In December 1971, as the storm subsided, *Mariner 9*'s systems were switched back on. Unlike its Russian counterparts, its computer was programmable after launch, and thus its mission could be altered as it went along. Such flexibility meant that this orbiter, of all the craft that had been launched that May, was the only one to succeed in its mission.

Mariner 9 approached Mars to 1,370 kilometers and began mapping the southern hemisphere from 25 degrees to 65 degrees south. It continued with up to 25 degrees of the northern hemisphere. By the time that it ran out of fuel on 27 October 1972, it had captured 7,239 stunning images of Mars—with sufficient resolution to reveal surface features as small as a football field.

Once again scientific concepts of our neighboring world were about to be turned on their head.

REVELATIONS

When the dust clouds subsided, they unveiled a Martian landscape that was a geologist's dream. Large inexplicable dark spots that had poked through the swirling storm clouds were disclosed as immense volcanoes—the gargantuan Olympus Mons, three times the height of Everest, and its fellows, Ascraeus Mons, Pavonis Mons, and Arsia Mons on the great Tharsis Bulge.

Scientists were awestruck by the Valles Marineris, the seven-kilometers-deep rift in the crust of Mars that stretches for a quarter of the planet's circumference, an amazing feature. Also revealed were the colossal impact basins of Hellas, Isidis, and Argyre—clues to the death of a once inhabitable world.

A once inhabitable world! For Mariner's cameras were the first to bring features to light that looked like dried-up riverbeds, valleys, and other telltale signs that large quantities of surface water—the prerequisite of life—had once been present here.

THE BECKONING PYRAMIDS OF MARS

On 8 February 1972, two months into its mission, *Mariner 9* passed over and photographed the area known as the Elysium Quadrangle. At 15 degrees north latitude and 198 degrees west longitude is a cluster of tetrahedral pyramidal forms, shown on frame MTVS 4205. This area was reimaged on 7 August, and frame MTVS 4296 showed the same area, once again with the pyramidal forms present.

These structures were first brought to scholarly attention in *Icarus* in 1974, in an article entitled "Pyramidal Structures of Mars." The authors noted that the structures cast regular shadows—showing that their tetrahedral forms are not illusions caused by albedo variations in surface soil coloration. The fact that there was more than one image taken at different sun angles further supports the view that their shape is not illusory.

These vast "beckoning pyramids," as Carl Sagan called them, tower a kilometer above the surrounding Elysium plain. It has been calculated that the volume of the largest is 1,000 times that of the Great Pyramid of Egypt, and that it is 10 times as high.

Are these features, as Sagan believed, "small mountains, sandblasted for ages"? He said that they warranted "a careful look."

WEIRD GEOLOGY?

There are four tetrahedral pyramids at Elysium—a larger and a smaller pair in close proximity, facing each other across the arid plain. They seem to have been set out in a definite pattern of alignment—a feature associated with pyramids on Earth—the two smaller pyramids seeming to mirror the alignment of the larger two.

Scientists have tried to explain them as wind-faceted volcanic cones, or as the result of peculiar forms of erosion or soil accumulation. But as J. J. Hurtak and Brian Crowley state in their book, *The Face on Mars:*

> This simple explanation does not stand up to closer examination. Wind-tunnel tests were done in Los Angeles in the mid-1970s by NASA engineers to simulate the creation of formations similar to those photographed by *Mariner 9.* All this experiment proved was that soil accumulation or wind sculpting would not provide for four equally

> spaced tetrahedral formations. It was not possible to simulate an evenly spaced arrangement of objects in the wind tunnel to match the mathematical distances one finds in the four major and minor pyramids in this area of Elysium.[5]

Other scientists have attributed these formations to glacial sculpting or eroded rotating lava blocks, but Hurtak and Crowley again disagree: "There is no evidence of glaciers [on Mars], especially within the tropic area of the planet [where Elysium lies] . . . and no lava spillage has been clearly detected in connection with the formations."[6]

What, then, are these enigmatic formations? Perhaps scientists have not been able to replicate them by simulating known types of natural processes because they were not produced by natural processes in the first place.

Could they be the first sign, as many independent researchers claim, that Mars is marked by the "fingerprints" of an ancient extraterrestrial civilization?

7

The *Viking* Enigma

THE next phase of Mars exploration came in 1975 when NASA launched the twin probes *Viking 1* and *Viking 2*. These craft were orbiter-landers like their ill-fated Soviet predecessors *Mars 2* and *Mars 3*. But unlike the Russian craft, the *Viking*s were to be an overwhelming success.

Viking 1 was the first probe to be launched, and on 20 July 1976 its lander module touched down safely on the Martian surface at Chryse Planitia, the great lowland basin lying to the north of the Valles Marineris.[1] Meanwhile, 2,000 kilometers above, the cameras in the orbiter had been switched on to acquire high-resolution photographs of the planet.

SEARCH FOR LIFE

Inspired by the revelations from *Mariner 9* that Mars could once have been habitable, NASA dedicated the *Viking* missions to the "search for life on Mars." For the most part this search was carried out by means of high-resolution photographs of large areas of the planet's surface, analysis of the structure and composition of the atmosphere, and chemical tests on soil samples gathered by the landers.

We saw in part 1 that the soil samples gave a number of positive results and that Dr. Gilbert Levin, one of the scientists who devised the experiments, remains convinced to this day that there is—at the very least—bacterial life on Mars. This is quite contrary to NASA's official view as it was recently put to us by Dr. Arden Albee, the project scientist for *Mars Global Surveyor:*

> I would say that none of the experiments indicated evidence of life. Several came out not exactly the way we expected because during the design of the instruments it wasn't understood that oxidants would be on the surface of Mars—and so they did not get results that were neat and clean as predicted, but they did not indicate the presence of life.[2]

CHOICE SITES?

Viking 1's lander had originally been scheduled to touch down on Independence Day, 4 July 1976, but the date was set back as scientists on Earth scanned live television pictures of the Martian surface transmitted by the orbiter. The preferred landing site looked dangerously rugged.[3] After some weeks of searching for a safer location, Chryse Planitia was chosen, and a successful landing was made there.

Now attention shifted to finding a suitable site for *Viking 2*'s lander. This is how Carl Sagan tells the story:

> The candidate landing latitude for *Viking 2* was 44 degrees north. The prime site, a locale named Cydonia, was chosen because, according to some theoretical arguments, there was a significant chance of small quantities of liquid water there, at least at some time during the Martian year. Since the *Viking* biology experiments were strongly oriented toward organisms that are comfortable in liquid water, some scientists held that the chance of *Viking* finding life would be substantially improved in Cydonia.[4]

Sagan and his colleagues were about to come literally face to face with something that looked very much like a sign of life—but not the kind of sign, nor the kind of life, they had imagined. Indeed, what they found was so beyond their comprehension that it was immediately dubbed an illusion *and was not allowed to influence the final choice of a landing site for* Viking 2.

ILLUSION

The discovery was made on 25 July 1976 by Tobias Owen, a member of the *Viking* imaging team at the Jet Propulsion Laboratory (JPL), Pasadena, California. He was examining frames of the Cydonia region for possible landing sites when he was heard to mutter, "Oh my God, look at this!"[5]

The frame that he was inspecting, reference number 35A72, showed an area of the Martian surface that was roughly split into two geological zones—an extensive plain, slightly cratered, with a handful of raised mesas, side by side with a rocky area of immense blocks of angled stone. Toward the center lay what appeared to be a gigantic humanoid face staring blankly up from the dead planet—serene, perhaps even imbued with pathos—a mute sentinel on the barren landscape.

Just hours later, Gerry Soffen, a spokesman for the *Viking* project, gave a briefing to the press about progress so far in NASA's self-proclaimed search for life on Mars. Somehow an image of the newly discovered Face had reached him, and he showed it to the journalists. "Isn't it peculiar what tricks of lighting and shadow can do," he commented dismissively. "When we took a picture a few hours later it all went away. It was just a trick, just the way the light fell on it."

Soon afterward JPL issued a press release making essentially the same points about the Face:

> Photo Caption: This picture is one of many taken in the northern latitudes of Mars by the *Viking 1* orbiter in search of a landing site for *Viking 2*.
>
> The picture shows eroded mesa-like landforms. The huge rock formation in the center, which resembles a human head, is formed by shadows giving the illusions of eyes, nose, and a mouth. The feature is 1.5 kilometers (1 mile) across, with the sun angle at approximately 20 degrees. The speckled appearance of the image is due to bit errors, emphasized by enlargement of the photo. The picture was taken on July 25 from a range of 1,873 kilometers (1,162 miles). *Viking 2* will arrive in Mars orbit next Saturday [August 7] with a landing scheduled for early September.[6]

UTOPIA

The next development was a decision from NASA that *Viking 2* would not, after all, land at Cydonia.

Apparently the site was now deemed "unsafe." According to Carl Sagan:

> 44 degrees north was completely inaccessible to radar site-certification; we had to accept a significant risk of failure with *Viking 2* if it was committed to high northern latitudes. . . . To improve the *Viking* options, additional landing sites, geologically very different from Chryse and Cydonia, were selected in the radar-certified region near 4 degrees south latitude.[7]

All this notwithstanding, it is an extraordinary fact that *Viking 2* was finally set down at a latitude even higher than Cydonia. It landed—and was almost overturned by boulders—on the distinctly unpromising rock-strewn plain called Utopia, at 47.7 degrees north latitude, on 3 September 1976. Thus—for no obvious reason says James Hurtak—"a multimillion-dollar effort may have overlooked 'paydirt' and may have become a trivial event. . . . A poor selective factor had been used to choose an area of minor geological and biological significance. It was like choosing the Sahara Desert as a suitable landing site on our own planet."[8]

THE LADY DOTH PROTEST TOO MUCH

Why choose Utopia over Cydonia when NASA's own criteria mark both sites as equally "unsafe," and when the former is bland and uninteresting while the latter has rumors of water and the mystery of the Face? The question is a nagging one, because even if we accept Gerry Soffen's instant dismissal of the Face as a trick of light and shadow, Cydonia still looks like a far more interesting site than Utopia.

Frankly we find the decision to land at Utopia baffling. But we are even more perplexed by the abrupt way that Cydonia was dropped as the preferred site so soon after the discovery of the Face on frame 35A72. It could be a coincidence. But on the other hand, we find it odd that NASA was in such a hurry to write off the Face as an illusion. In a way spokesman

Gerry Soffen was perfectly correct to state that the image vanished within a few hours. This did not happen, however, because of tricks of light and shadow, but because night had fallen. *No image of the Face was acquired a few hours later.*

Quite simply, the much-vaunted photograph that proves the Face is an illusion does not exist.

So why, then, did NASA spread this strange story around?

8

Jesus in a Tortilla

On 4 July 1997, *Pathfinder,* the first of a new generation of NASA probes, landed on the rust-red surface of Mars at Ares Vallis (19.5 N, 32.8 W), bounced in its protective gas-filled airbags, and came to rest intact on an alien world.[1] Then, as though in a scene from a science-fiction movie, the airbags deflated and three triangular solar panels opened like the petals of a futuristic silver flower. A ramp rolled out and the *Sojourner* rover was deployed. The world watched in awe as this tiny six-wheeled robot, the size of a shoe box and just 10.5 kilograms in weight, crept out from its protective metal flower and edged onto the Martian soil to find itself marooned on that rock-strewn world, under a salmon pink sky—millions of miles from home.

MARS OBSERVER, PLEASE PHONE HOME

Pathfinder was hailed as a roaring success by all those involved on the project. NASA could now breathe a sigh of relief after the patchy record of the previous decade, which had started with the horrendous inflight explosion of the space shuttle *Challenger* in 1987 and had included the loss in 1993 of the Mars probe *Mars Observer.*

Launched on 25 September 1992, *Observer*'s mission was to re-map the surface of Mars—essentially duplicating the photographic work of the *Viking* orbiters, but at much higher levels of resolution. It carried a camera that could obtain images at 1.4 meters per pixel—a vast improvement on the 50 meters per pixel for which the *Viking*s were capable.

But *Observer* failed just before going into orbit. A NASA press release describes what happened:

> On Saturday evening, August 21st [1993], communications were lost with the *Mars Observer* spacecraft as it neared to within three days of the planet Mars. Engineers and mission controllers at NASA's Jet Propulsion Laboratory, Pasadena, California, responded with a series of backup commands to turn on the spacecraft's transmitter and to point the spacecraft's antennas toward Earth. As of 11:00 A.M. EDT on Sunday, August 22nd, no signal from the spacecraft had been received from tracking stations around the world.[2]

CONSPIRACY THEORIES

What exactly happened to *Mars Observer*?

Though there was almost no specific evidence on which to make judgments, an independent NASA review board was set up to answer this question. After deliberations the board suggested that a rupture in a line in the propulsion system during the start of fuel-tank pressurization somehow blacked out the spacecraft's communications with base.

But there was more to it than that, and a few days later it became clear that there had been a huge breach of procedure. What had really happened was that *Observer*'s radio link (telemetry) to Earth had been *deliberately* shut off by the controllers during the period that the fuel tanks were pressurizing. This was bizarre and unprecedented. They must have known how vital it is that communication between spacecraft and base should be maintained at all times—once lost it is hard to retrieve. This is precisely what happened to *Observer:* having been cut off, its telemetry could not later be reestablished.

At the very least the loss of the probe was stupid. But as we report in chapter 15, some NASA analysts were convinced from the beginning that there may be more to it than that. They point out that *Observer* was supposedly ready to start its mapping orbit when the telemetry was shut down. Why, they ask, would such a risky procedure even have been contemplated at such a crucial juncture—unless NASA had actually *wanted* to lose the spacecraft.

The motive?

Conspiracy theorists are convinced that the whole mystery is connected to the growing publicity around the issue of the Face during the decade prior to *Mars Observer.* After all, in the run-up to the September 1992 launch, there had been vociferous public demands that the probe should rephotograph Cydonia.[3]

Maybe it went into orbit a few days earlier than the public were told? Maybe it did photograph Cydonia? Maybe the powers in NASA didn't like what they saw there? Maybe they decided to pull the plug not wishing to disclose to the volatile masses the potentially disturbing news of the reality of extraterrestrial life?

DIPIETRO, MOLENAAR, HOAGLAND

NASA has done much to fuel such paranoia by dissembling about the Face since the moment Tobias Owen first spotted it in *Viking* frame 35A72 on 25 July 1976. Cleverly worded snippets of official disinformation fixed it in the public imagination as nothing more than an illusion of light and shadow. Scientists en masse instantly lost interest in it. And for the next three years it lay buried in NASA's deep-space archive at Goddard Space Flight Center, in Greenbelt, Maryland.

The Face was rediscovered in 1979 by Vincent DiPietro, a Lockheed computer scientist on contract at Goddard. Working with his colleague Gregory Molenaar, he developed a process of image enhancement to create more detailed images of the Face. On their own initiative, as we shall see in chapter 9, the two researchers also combed the archives and found another *Viking* frame in which the Face, although photographed from a different angle, was clearly visible. In this frame a second enigmatic structure could also just be made out—a mysterious five-sided pyramid (subsequently named the D&M Pyramid) within 10 miles of the Face.

DiPietro and Molenaar at first naively supposed that NASA would be interested in their discoveries. Predictably, they were soon disappointed. Here were two scientists, employed by NASA, holding immaculate qualifications, who were effectively claiming that they had found evidence of intelligent design on another world. Yet no one would listen to them.

In 1981 they gave up trying to push the matter through official channels and published their own book, entitled *Unusual Mars Surface Features.* Among those who picked up a copy at the launch party was a science writer, Richard Hoagland, who by coincidence had been among the gaggle of press members at JPL in July 1976 in whose presence Gerry Soffen had so glibly explained away the Face.

Hoagland, a veritable jack of all trades in the scientific and space world with a prodigious CV, would become, in time, the main publicist and controversial figurehead of the early Cydonia researchers. Referred to by his own editor as "a curious combination of *Star Trek* creator Gene Rodenberry and Mr. Spock,"[4] this maverick was to bring DiPietro and Molenaar's discoveries into the public eye—and in the pre-millennium zeitgeist there was a ready audience interested in such a stark challenge to conventional scientific thought.

INDEPENDENT MARS INVESTIGATION

As well as stirring up a storm of publicity, Richard Hoagland made a number of pioneering discoveries of his own among the *Viking* frames. These included what he termed the "City," the "Fort," and many small mounds within a few miles of both the D&M Pyramid and the Face.

With anthropologist Randolpho Pozos, Hoagland established the Independent Mars Investigation (IMI) in 1983. They set up a computer conference—named after the Ray Bradbury book *The Martian Chronicles*—in which Hoagland, Pozos, DiPietro, and Molenaar were joined by plasma physicist John Brandenberg and artist Jim Channon (who would provide an artistic evaluation of the Face). Other members of the conference included Lambert Dolphin and Bill Beatty, both scientists from the Stanford Research Institute (SRI), the world-famous California think tank. Dolphin, a physicist, had for some time been involved with remote sensing surveys around the pyramids and the Great Sphinx on Egypt's Giza plateau.

The Independent Mars Investigation was taken seriously enough to be granted $50,000 from the President's Fund at SRI—though it soon became apparent that the think tank did not want to give further assistance, allowing only Dolphin's spare time and some technical support. Moreover, even this limited backing looked as though it might at any time be withdrawn. In desperation Hoagland formed a second group—the Mars Investigation Group, with Thomas Rautenberg, of Berkeley, California. Meanwhile, in March 1984 the IMI conference folded and the Martian Chronicles came to an abrupt end.

IMI's main conclusions were presented by John Brandenburg at the Case for Mars Conference II held at Boulder, Colorado, in the summer of 1984.

CARLOTTO

In 1985 the independent researchers were joined by a computer programmer, Mark Carlotto, who was a specialist in imaging techniques. As we shall see in chapter 10, Carlotto worked on the original *Viking* images, enhancing them and finally concluding that the Face is a three-dimensional object with many characteristics that appear to be artificial.

Carlotto is an impressively qualified scientist, and his work has never been anything other than scientifically rigorous. Nevertheless, he was to find that his conclusions and observations were, from the outset, utterly rejected by Mars experts.

THE McDANIEL REPORT

Some academics from other disciplines who have looked into the findings of independent scientists such as Carlotto, DiPietro, and Molenaar believe that the "expert" reaction to them has been ill-considered.

For example, Stanley McDaniel is professor emeritus and former chairman of the Department of Philosophy at Sonoma State University. He first heard about the Face controversy as early as 1987. In 1992, spurred on by the impending launch of *Mars Observer,* he began his own independent evaluation of the Cydonia debate:

> My initial approach was one of considerable skepticism . . . but over the course of the investigation my appreciation of what the researchers had done, and the underlying scientific integrity of their work, began to grow.

> I found that the occasional faults in their work were far outweighed by the solidity of the data and their responsiveness to the needs of what is, after all, the first study of its kind in history.
>
> I became aware not only of the relatively high quality of the independent research, but also of glaring mistakes in the arguments used by NASA to reject this research. With each new NASA document I encountered, I became more and more appalled by the impossibly bad quality of the reasoning used. It grew more and more difficult to believe that educated scientists could engage in such faulty reasoning unless they were following some sort of hidden agenda aimed at suppressing the true nature of the data.[5]

A slight, energetic man, Stan McDaniel is a brilliant orator and a quick thinker—a personal affront to the theory that the "Artificial Origins at Cydonia" (AOC) hypothesis is only supported by "unscientific" types. The subtitle of his report, which was published in 1993, spells out its central conclusions: "The failure of executive, congressional and scientific responsibility in investigating possible evidence of artificial structures on the surface of Mars, and in setting priorities for NASA's Mars exploration program."[6]

The McDaniel Report sets out to analyze not only the artificiality argument, but also NASA's countermeasures against it. Foremost among these is the standard defense—much promoted by Carl Sagan—that the Face is just a trick of light and shadow. Then there is a so-called technical report (McDaniel claims it is nothing of the sort) that criticizes Hoagland's *Monuments of Mars,* as well as the work of Dr. Michael Malin, the designer and operator of the cameras carried by the probes. A staunch opponent of artificiality, Malin holds the power to choose what on Mars will be photographed on any mission involving his cameras as well as a strange legal privilege—a six-month "probationary" period during which he is permitted to view the images before they are released to the general public.[7]

There can be little doubt that Carl Sagan, while he lived, was an extremely effective NASA spin doctor calming public concerns about the Face. He even wrote a piece on the subject for the Sunday newspaper magazine *Parade* in which he staunchly defended NASA's "illusion" arguments about the Face and likened it to many faces that appear in nature, such as the Great Indian Face, the Man in the Moon, and Jesus in a Tortilla.

It is with exactly such arguments that NASA has consistently defended its policy of not prioritizing Cydonia. But are its arguments really valid—or merely dismissive? In McDaniel's view they are the latter. Indeed, they are not only dismissive but fundamentally flawed, perhaps even deliberately flawed.

LOST PROBES

Mars Observer offered the ultimate means to settle the controversy—new high-resolution photographs of the Cydonia plains—but only if NASA and Michael Malin could be persuaded that it was worth pointing *Observer*'s camera in the right direction. The lobbying began in earnest. Then, just twenty-four hours before Richard Hoagland was scheduled to debate the matter live on national TV with *Observer* scientist Bevan French, the probe was lost.

It was not the first probe in recent history to have been mysteriously silenced. Two Russian probes sent to Mars in 1988 also lost contact. *Phobos 1,* launched on 7 July 1988, was deemed lost after just 53 days, while *Phobos 2,* launched three days after *Phobos 1,* managed, it is thought, to map some of Mars. It was somehow destroyed while imaging Phobos, one of the tiny moons of Mars. The last image it sent back to Earth was of a huge baffling cigar-shaped elliptical shadow—miles long—on the Martian surface.[8]

GLOBAL SURVEYOR

As we write these words, *Mars Global Surveyor*—the successor to the doomed *Mars Observer*—is engaged successfully in the mission; its predecessor failed even to begin.

Essentially it is a less expensive *Observer,* with only five of the original seven experiments on board, yet it still has the same Malin Space Science

Systems Camera, and Dr. Malin still presides over the use of this piece of modern technology.

But what of NASA's official policy? Is it the same as before? Has the work of the AOC researchers convinced them to make a thorough study of Cydonia?

Face Staring Back

Oh! I have slipped the surly bonds of Earth,/And danced the skies on laughter-silvered wings; . . . /Up, up the long, delirious burning blue/I've topped the windswept heights with easy grace,/Where never lark, or even eagle flew/And, while with silent lifting mind I've trod/The high untrespassed sanctity of space,/Put out my hand and touched the face of God.

JOHN GILLESPIE MAGEE, JR., "HIGH FLIGHT," 1943

A photograph is not only an image (as a painting is an image), an interpretation of the real; it is also a trace, something directly stenciled off the real, like a footprint, or a death mask.

SUSAN SONTAG, *NEW YORK REVIEW OF BOOKS*, 23 JUNE 1977

WHEN Tobias Owen discovered the Face on Mars in *Viking* frame 35A72 he reacted in a totally natural way: "Oh my God, that looks like a face."

Typically the image does produce this response—an instantaneous gut reaction of recognition. But is it really what it seems? Or is it just a trick of light and shadow? Some very intelligent and highly qualified people have spent a great deal of time during the past twenty years trying to answer these questions.

SECRETS OF PIXELS

Vincent DiPietro, the first scientist to take the Face seriously (and the man who rediscovered it in the Goddard archives in 1979) is an electrical engineer, specializing in digital electronics and image processing. He shared the discovery with fellow Lockheed computer scientist Gregory Molenaar, who was on contract to NASA with the Computer Sciences Corporation and has a similar background in computer image analysis. Seeing the whole process as an "adventure," the pair embarked on a clandestine project to enhance the image of the Face and to reexamine the original Viking data tapes for other anomalous objects on the Martian surface.[1]

The Face occupies an area of only 64 × 64 pixels in the original image, with each pixel representing an area of 150 × 155 feet.[2] *Anything smaller than this simply does not register.* Nevertheless the pixels are encoded with helpful clues that enable computers to reconstruct what is there.

As the orbital camera was of low resolution, it had to average out the tone of each 150 × 155–foot area to come to a value for the pixel that would represent it. To the lightest areas it assigned a low numerical value (white = 0) and to the darkest areas it assigned a high value (black = 256). The orbiter was then able to transmit the images back to earth as a sequence of numbers that could be printed out as black-and-white pictures built up out of "gray-scale" pixels.

The image-enhancement work done by DiPietro and Molenaar was an attempt to glean some detail from each pixel about what lay below its average 256 tone. This could be done by comparing each pixel with its neighbor. For example, if one pixel was light gray and its neighbor on the left lighter, and its neighbor on the right darker, it was probable that these three blocks of tone actually represented a gradual change from light to dark, not a markedly stepped difference in tone, from left to right.[3] Using such an approach, more detail could theoretically be squeezed out of the grainy *Viking* images:

> In order to magnify digital images, additional pixels must be added and their values determined. [One] method is to calculate intermediate pixel values . . . using some combination of their surrounding values. For example, bilinear interpolation uses a pixel's four nearest neighbors and produces results that are smoother than pixel replication, but tend to be blurry.[4]

SPITTING IMAGE

The first step was to clean up frame 35A72 by removing transmission errors (errors due to interference, etc., characterized by pure white or black single pixels). Next, realizing that most of the data on the frame was between gray-scale values 60 and 108, DiPietro and Molenaar stretched the contrast so that 60, not 0, became white and 108 became black. Thus the middling gray tones of which the images were made were replaced by a broader range of light and dark.

This was better but the researchers were still not satisfied with the results, which they described as "huge pixels with stair-step-like images." They therefore "designed a way to remove the ragged edges by dividing each of the original pixels into nine smaller units. Each new pixel is shaded by summing percentages of the original adjacent pixels with the subject pixel to aim at discreet new values."[5]

They named this process SPIT, after "spitting image," and the acronym for Starburst Pixel Interleaving Technique. As a control they subjected terrestrial low-resolution satellite photographs of the Pentagon and Dulles International Airport in Virginia to SPIT processing and achieved much clearer images, which were verified against aerial photographs of the sites.

Satisfied that their technique worked, DiPietro and Molenaar now used it on frame 35A72:

> A remarkable improvement occurred. The Face began to reveal much more detail than had previously been observed.[6]

MISSING FRAMES

In 1976 NASA spokesman Gerry Soffen had stated categorically that another image of Cydonia—on which the Face "disappeared" in a different sun angle—had been acquired just "a few hours later" than frame 35A72. Naturally DiPietro and Molenaar wanted to study this frame, but an exhaustive search proved it did not exist in the archives. Indeed, as we have seen, Soffen was being either presumptuous or economical with the truth when he made his 1976 statement—for "a few hours later" Cydonia had been in darkness and the *Viking* orbiter had been elsewhere, photographing an entirely different part of the planet.

The two Lockheed scientists persevered, however, and eventually did come across one other Cydonia frame showing the Face—frame 70A13. It had been acquired thirty-five days later than 35A72 and had been curiously misfiled. *This is the only other frame that shows the Face.* When it was taken the sun was much higher than it had been on frame 35A72 (27 degrees instead of 10 degrees). Far from "disappearing" under this different sun angle, the Face was still clearly visible:

> Not only did the second frame confirm the first, but additional features emerged. The contour of the eye cavity remained unchanged. The second eye cavity became more distinct. The hairline continued to the opposite side. A chin line began to take shape.[7]

Next DiPietro and Molenaar replaced the gray-scale tonal values in the the two frames with a scale based on colors. They did this because color differences are easier to see than shades of gray. The result was that the contents of the eye cavity began to become visible. To their amazement the researchers found themselves looking at something very much like a representation of an eyeball with a distinct pupil.

This, then, was the initial evidence put forward by DiPietro and Molenaar—strongly suggesting that there was much more to the Face than a trick of light and shadow. But were they right?

Before coming to any conclusions of our own on the matter we felt we needed a second opinion on the imaging techniques DiPietro and Molenaar used.

AN EXCITED DR. WILLIAMS

A good place to start asking questions was NASA itself, with the scientists currently working on the *Pathfinder* and *Global Surveyor* missions to Mars. In July 1997, therefore, three weeks after *Pathfinder* had touched down in Ares Vallis, we arranged a meeting with Dr. David Williams, *Pathfinder*'s chief archivist at the Space Science Data Center in the Goddard Space Flight Center, where DiPietro had rediscovered frame 35A72.

Goddard is a huge expanse of offices and laboratories set in lush Maryland countryside half an hour by car from the center of Washington, D.C.

Feeling a little daunted by the military thoroughness of the security procedures, we picked up our passes at the gatehouse and were ushered inside.

After a ten-minute walk along a pleasant wooded road we reached the archives building. Expecting to find a grizzled, die-hard scientist, we were pleasantly surprised at Dr. Williams's youth and enthusiasm, which sharply contrasted with NASA's official image. Better still, Dr. Williams was keen to talk about the Face on Mars:

> Well, I know that there's a number of scientists, serious scientists, who are working on this from the angle that it's an artificial structure—a sign of intelligence—so personally I would like to see what *Mars Global Surveyor* finds when it takes its images, hopefully high-resolution, different lighting angles, things like that, to see what this area looks like, what this "face" looks like.
>
> I would be surprised if it did not turn out to be natural, but on the other hand, I think it would be pretty cool if it wasn't! That would be neat, imagine it—if pictures came back and unequivocally said this was an artificial structure. I mean, it would change our whole view of the entire universe. I think that would be pretty exciting.

NEW FOR OLD

As chief archivist for the *Pathfinder* mission, Dr. Williams has to assess and interpret incoming data. He was therefore the appropriate person to give us NASA's views about the nature and validity of the enhancement techniques used on the earlier *Viking* images.

Only the raw *Viking* images could strictly be said to be 100 percent accurate, he pointed out. But, he admitted, it is standard practice at NASA to manipulate such images to make them cleaner and more defined:

> If you open up a raw *Viking* image, most of them look like there's nothing there, and even though it doesn't take long, you have to enhance the contrast, you have to stretch it, you have to do things so that you can actually see what is really in the image.

Indeed, he confirmed, the computer enhancement of received raw data is not only standard procedure but is absolutely necessary to make sense of the kind of information transmitted by orbiting cameras. He also confirmed that techniques such as the SPIT process devised by DiPietro and Molenaar are now used in a great many commercial applications. As he pointed out, DiPietro and Molenaar had recently received an award from the Computer Sciences Corporation of Virginia for developing the SPIT process, which has proved itself as an effective method for extracting information from computer images.

ARTISTIC MERIT?

In the early days of his research, Richard Hoagland suggested that artists should evaluate the ratios and proportions of the Face. He reasoned that if it accorded with artistic criteria then this would be another sign of artificiality. Jim Channon, artist, concept designer, and illustrator, took up the challenge.

Channon concentrated on proportions (anthropometry), the supporting structure (architectural symmetry), and expression (artistic cultural focus). His conclusions were as follows:

> I find no facial features that seem to violate classical conventions. The platform supporting the Face has its own set of classical proportions as well. . . . Were the Face not present, we would still see four sets of parallel lines circumscribing four sloped areas of equal size. Having these four equally proportioned sides at right angles to each other creates a symmetrical geometric rectangle. These support structures alone suggest a piece of consciously designed architecture.
>
> The expression of the Face on Mars reflects permanence, strength, and similar characteristics in this range of reverence and respect. There is overwhelming evidence that the structure revealed in the photographs presented to me by Dick Hoagland is a consciously created monument typical of the archaeology left to us by our predecessors. I would need much more precise evidence at this point to prove the contrary.[8]

NEW FEATURES

Channon's analysis was done before computer analyst Mark Carlotto had re-imaged the *Viking* frames using techniques that improved upon those of DiPietro and Molenaar. We will review Carlotto's work in more detail in chapter 10. But, briefly, what it revealed was a highly controversial set of new features on the Face—features that would echo, as Channon had said, monuments "typical of the archaeology left to us by our predecessors." These features include teeth, a diadem, a teardrop, and a distinctive headpiece decoration that is striped like the characteristic *nemes* headdress worn by the pharaohs of Egypt (seen as the headdress of the Great Sphinx of Giza).

Carlotto's work on the second frame, 70A13, revealed that the Face is not as symmetrical as other researchers had previously thought. Using a technique known as "cubic spine interpolation," which greatly enhances contrast, he was able to pick out details in the Face that previously had been too faint to be noticed.

Its left side, in shadow on frame 35A72, is better lit on frame 70A13, which was taken at a higher sun angle. The left eye socket can be seen and the mouth is revealed as not quite straight. Instead it seems to rise upward at the corners, as though in a sneer.

Carlotto also uncovered a "convoluted" area below the left cheek. Some see it as a kind of ramp, but this is pure speculation because the relevant area is marred by either a crater or a camera registration mark that cannot be removed from the enhancements.

A "TRICK OF LIGHT AND SHADOWS"

On 31 July 1997, twenty-one years after NASA's first attempt to explain the image of the Cydonia Face as an illusion, we traveled to Pasadena, California, to visit the California Institute of Technology. This private university and think tank runs NASA's nearby Jet Propulsion Laboratory and has been home to some of the legendary scientists of the century, including the Nobel Prize–winning physicists Albert Einstein and Richard Feynman.

The impeccable buildings of Cal Tech nestled beneath the San Gabriel Mountains spread out among lush gardens and cooling fountains. Unlike the anonymous, heavily guarded blocks at JPL, one can roam the aesthetic vistas of Cal Tech at ease. We found refuge from the burning heat in the air-conditioned office of Dr. Arden Albee.

We were lucky to see him. After spending hours on the phone, being passed from pillar to post, we were finally, in desperation, put through to him. He was leaving for Japan the next day to discuss his work as chief scientist on the *Mars Global Surveyor* mission, which was then fast approaching Mars orbit. This craft was destined to re-image the whole of the Martian surface—including the Cydonia region. On the eve of a possible test of the Artificial Origins at Cydonia hypothesis, what did *Mars Global Surveyor*'s chief scientist, and onetime JPL chief scientist, make of the furor?

Dr. Albee is a busy man, at a busy moment in Mars research, and we were grateful for his time. Replying slowly, with deliberate emphasis, he answered our queries as though he was at one of the numerous press conferences that had been a common event for him in the preceding weeks. At the mention of Cydonia his face dropped. What, we asked, was his opinion of the Face on Mars and the case for its artificiality made by AOC researchers?

> What it is is a shadow that has an appearance that somewhat resembles a face. And so there is a difference in the albedo [surface coloration], in that pixel by pixel the return clearly has some resemblance to a face, and what their [the AOC researchers] calculations did was to assume that these differences in color or differences in albedo, really, were due to slope—because that's how your eye looks at it, and sort of says, hey, that's a slope! It doesn't have to be that, it could be changes in the amount of dust on the surface; it could be partly slope, partly dust, partly different material, and so on. It is a trick of light and shadows.

We asked Dr. Albee if he knew of the *McDaniel Report,* or the work of DiPietro, Molenaar, Hoagland, or Carlotto. In answer, with a broad grin, he took down a copy of the *McDaniel Report* from his bookshelf.

> You know, people dream up all kinds of crazy things. Every place you go there's a tourist spot, whether it's in the Alps or Wisconsin, or the Grand Canyon—the great Indian face or the great Yogi bear. People look at natural things and see human faces in them. It's a natural phenomenon, it goes back to prehistory.

"IS THAT A CAMEL?"

Following the Arab uprising of 1917, T. E. Lawrence (Lawrence of Arabia) presented the leaders of the rebellion with portraits of themselves. To his amazement they literally could not see what the paintings were supposed to be. One tentatively pointed to the image of his own nose and asked, "Is that a camel?"

The Arabs were not being ignorant and naive. They merely lacked the specifically European cultural references of the time that would have taught them to know what to look for. All they could see was a flat square canvas covered in areas by colored paint. They were at first unable to interpret these areas of pigment as representations of three-dimensional objects. In a way they were seeing reality, and it is we who are taken in by illusion. What the Arabs saw was what was actually there. They were unaware that a picture is a visual metaphor. We, however, would have seen a face—where there really was, in truth, only pigment.

In the same way, as you read these words, the sounds that you hear in your head are not intrinsic to the printed letters. An alien, on seeing this page, would just see it as a mass of squiggles—and again, like the Arab chiefs, would be correct. It is we who are culturally educated to transmute the shapes into sounds—which, of course, they are not.

Recognition of faces as significant objects is a genetic predisposition of the human species, something that we inherit and never need to be taught—indeed, something that is hard-wired into the brain itself.[9] Obviously it is an important gift. It means, for example, that a newborn will instantly recognize human beings (preferably its parents) without first having to learn what humans look like.[10] Thus it is that any arrangement of objects that resemble facial features, whether they be a face or not (they could be two apples, a carrot, and a banana), will act as a stimulus to the brain and cause us to see that object, or collection of objects, as a face. For the same reason, we sometimes see faces in clouds or become scared of a tree that seems to have a twisted evil face in its bark.

But face recognition is not quite the same skill as recognition of an *image* of a face. As Lawrence's example shows, the ability to see a face in a two-dimensional portrayal such as a painting or a photograph is something that has to be learned. Had the Arabs been given sculptures, there is no doubt they would have seen that they represented faces.

For the sake of argument, let us imagine that the *Viking 1* orbiter that photographed Cydonia was not an unmanned mission but was crewed by the 1976 equivalents of T. E. Lawrence and one of his Arab allies.

Drifting some 1,800 kilometers above the surface of the Red Planet, armed with a powerful telescope, our two protagonists would pass over the Face and exchange observations. Lawrence would turn to his colleague and say, "Wow! Look at that face!" But what would the Arab say? This is the question that goes to the very heart of the Artificial Origins at Cydonia hypothesis. Is the Face merely an illusion, a Rorschach image, on which Lawrence is projecting qualities that do not belong to it—and which the Arab cannot see except as a two-dimensional pattern of differing tone values? Or is the object truly sculpted (by nature or artificial means), in which case the Arab sees it? Does he reply "What face?"—or does he too gasp in wonder at the dusty visage staring back up at him?

10

Ozymandias

MARK Carlotto of the U.S. Analytic Sciences Corporation is a major figure in the debate over artificial origins at Cydonia. Since first hearing about the Face on Mars in 1985 he has consistently been at the forefront of research, using his skills as an image processor to extract new, high-quality information from the original *Viking* data tapes. He told us in an interview in December 1996:

> My initial reaction was kind of open-minded; I was intrigued. I had no idea about this. I've always followed the space program pretty closely, since college, and I was in college in 1976. I remember *Viking*—but I didn't hear then about the Face on Mars. So I was curious. . . .
>
> I started off applying the methods that we used in my day job at the Analytic Science Corporation, TASC. What I did was to apply the methods we were using routinely at the time to enhance X-rays, radiographic analysis, remote sensing, satellite images, that sort of thing. I was really able to clean up and restore the [original *Viking*] imagery.[1]

THREE-DIMENSIONAL ANALYSIS

We have spoken of Carlotto's images in previous chapters and noted that they show intriguing features and previously unnoticed details in the Face; for example, bilateral crossing lines above the eyes that are suggestive of a diadem, teeth in the mouth, and stripes on the headdress. Carlotto was also able to add to the stock of information on other previously known attributes of the Face such as the left eye socket (on the shadowed side) and an alleged teardrop below the right eye.

"I was bothered from the start," he told us, "by NASA's 'trick of light and shadow' hypothesis. And so I figured, well, maybe there's a way of assessing this, and that's when I got into three-dimensional analysis of the Face to reconstruct its shape and get an awful lot of details a lot more clearly."

Such analysis gleans information about the three-dimensional aspects of an object from its two-dimensional representation; that is, a photograph. This can be done in a variety of ways depending on the available imagery: by the analysis of the heights of the shadows; by stereoscopy—comparing two images of the same object taken from different angles; and in particular by "shape from shading," also known as photoclinometry.[2] As Carlotto puts it: "Shape-from-shading techniques reconstruct the shape of the object being imaged by relating shading information to surface orientation. In cases [such as Cydonia] where there is a lack of distinct surface features and texture, the primary source of surface information is shading."[3]

One objection to shape from shading is that the computer may end up doing exactly the same job as the human brain. In other words, it may "see" shade as slope—for example, interpreting what could be nothing more than flat-surface albedo coloration as height. The great strength of the computer, however, is that it can build 3-D images and then view and test these from different angles and perspectives.

Working with the two available *Viking* frames of the Face, Carlotto instructed his computer to prepare three-dimensional models based on each of them. Since the two frames had been taken at different angles and different times of day he wanted to see if the computer would construct different models from each. Both reconstructions showed facial features in the underlying topography, however, an indication that the structure is indeed three-dimensional and face-like.

Carlotto then checked his results in an ingenious fashion. Taking the model of the Face from frame 35A72, he instructed the computer to illuminate it from the sun angle given in frame 70A13. His image correctly predicted the shadowing that was found on the real 70A13. He then repeated the procedure, this time using the sun angle from frame 35A72 on the photoclinometrically reconstructed face from frame 70A13. Again the computer image paralleled the real frame.

FRACTALS ON MARS

Most of mankind's giant leaps forward in space discovery have followed advances in weapons technology. It should therefore come as no surprise that the computer-processing technique best adapted to detecting signs of artificiality in the Cydonia images is one that was originally developed for military purposes. "At the Analytic Sciences Corporation," Carlotto told us, "we were at that time developing computer programs for detecting manmade objects. Again, I went into the analysis with an open mind. I simply took the technique we were using on terrestrial imagery and applied it to the Mars imagery, right down to the same settings and everything."

The programs that Carlotto was developing for TASC involved what is known as "fractal analysis." Put simply, nature tends to repeat herself in specific areas in terms of the morphology of natural features. An example is the fronds of a fern—each of which is a scale model of the larger, whole fern—or cracks in rock, which resemble great mountain crevices, only on a smaller scale. The basic patterns that make up natural structures are termed fractals, which are repeated on a range of different scales. Because of this quality of natural objects to be self-similar, a computer can be used to detect the repetition of the basic morphological fractal and thus distinguish a natural object from an object that does not correspond to the fractal pattern—i.e., an object that is almost certainly artificial.

For the military, this technique can be used to detect manmade objects and installations camouflaged in any terrain. First the computer calculates the "normal" fractal model for the locality, then it analyzes the entire region and highlights any parts of that terrain that do not seem to fit the fractal model. If these objects are non-fractal to any great degree, then they are judged alien to that specific locality; that is, they are in all probability manmade. It has been calculated that fractal analysis correctly identifies artificial objects with roughly 80 percent accuracy.[4]

Carlotto with a colleague, Michael C. Stein, carried out a detailed fractal analysis of the *Viking* frames:

> We found that the Face was the least natural object on frame 35A72 and applied it to adjacent frames. It was also the least natural object over the four, five frames that we did. Very anomalous.[5]

In fact, Carlotto's fractal analysis revealed the Face as the least natural object for 15,000 kilometers in every direction—showing a model-fit error curve slightly more pronounced than that of a military vehicle!

ILLUMINATION

Whatever it may finally prove to be—artificial work of sculpture or weirdly eroded mesa—the Face on Mars is *not* a "shadow that somewhat resembles a face." It looks like a face, because it is face-like in form. We believe that Carlotto's work proves this. But it does not prove artificiality—in part because the unilluminated side of the Face is in general much less convincing than the illuminated side, as Carlotto readily admits:

> It is apparent that the shadowed side of the Face is either incomplete or degraded and is not a mirror image of the side in sunlight. Those who support the intelligence hypothesis argue that the distortion could be due to meteorite impact, erosion over time, outright abandonment of the project, or its intentional discontinuation upon achieving adequate recognizability as a face. Opponents are not surprised at the roughness in the symmetry of what they believe simply to be a naturally formed mesa.
>
> It should be understood by all concerned that the original *Viking* data from the shadowed side of the Face contains very little information and therefore represents the weakest link in the chain of image reconstruction. Final judgments about the symmetry of the ridge line and the nature of any fine detail in the shadowed side should be suspended until the Face can be photographed under more revealing illumination.[6]

On 5 April 1998 *Mars Global Surveyor* did succeed in rephotographing the Face under more revealing illumination and in high resolution. As we shall see in chapter 15, the image remains ambiguous. Yet the Face does not stand alone and, as Carlotto told us when we interviewed him in December 1996, it is the context in which the Face is set that provides the most convincing evidence of artificiality:

> About a year ago I began to see another direction here, another avenue of research. Coincidentally, over these last few years I'd been getting increasingly more involved in "Bayesian analysis"—this is a way of really taking lots of pieces of evidence and putting them together and qualifying to what extent these support or deny your hypothesis. The thought occurred to me about a year ago, maybe this could be applied to include all the evidence about [artificiality at Cydonia], not only the work that I have done, but also the early discoveries of Hoagland and others.
>
> So during this last year I think I've been transformed in some sense, in that when I first got involved in this I was open-minded, but I wasn't ready to jump on the bandwagon and wave a flag. I've always been very cautious. . . . Up until a year ago if someone would ask me, "What do you think the odds are [of the Cydonia structures being artificial]?" I would say, "51 percent to 49 percent"—a real conservative engineering kind of assessment. But, I've always been split-brained on this. . . . I guess intuitively I felt there was more there, but it was subliminal. This Bayesian analysis, I think has, in my mind, just brought it out that there's no single piece, no "smoking gun." Instead there's a lot of little pieces that all kind of add up. . . . At this point in time I feel pretty confident these are artificial objects.

LOOK ON MY WORKS . . .

Inspired by the ruins of the giant statues of Ramses II on the west bank of the Nile at Luxor, Percy Bysshe Shelley (1792–1822) wrote "Ozymandias," his haunting poem of hubris and destruction. It tells of a traveler coming on the ruins of the vast, broken statue of Ozymandias, King of Kings, on which he reads, "Look on my works, ye Mighty, and despair." The king, in his pride, wants readers to look at the splendid city that he rules over, wishing them to despair in the face of his power, but time has reduced his works to dust. The meaning of the line therefore twists into a warning of mortality to those proud rulers like Ozymandias who think themselves mightier than death.

Were we to stand on the Cydonian plain, we, too, would see a "half-sunk, shattered visage" in the sand. From this proximity we could tell if we were looking at just a hill, or whether we were dwarfed by the crumbling death mask of some ancient alien Ozymandias.

Perhaps we could even look upon his "works"? For if we were to cross the once-flooded plain to the foothills of the ancient shoreline, we would come to a place where a city, though in ruins, may still stand.

11

Companions of the Face

THE Face is not alone on the plains of Cydonia but is surrounded by other anomalous structures which some believe will ultimately prove to be of greater importance. Richard Hoagland has even suggested:

> If someone made it with the purpose of attracting our attention, there was a certain logic to a face. What better way to call attention to a specific place on Mars as a site for further exploration?[1]

Hoagland had been present at JPL on the day the Face had been discovered in 1976, and like the rest of his colleagues in the press, had initially believed Gerry Soffen's "illusion" explanation. Only years later, with time to peruse the image in more detail, did he get bitten by what he calls "this Mars bug." He remembered a facetious comment that had been made on that afternoon at JPL by a fellow journalist—along the lines of "the Face is to tell us where to land." Ignoring the intended sarcasm, Hoagland decided to take seriously the possibility that the Face could be a marker for something else and began to search the Cydonian landscape for other "monuments."

THE CITY AND THE FORT

Reasoning that whoever had created the Face would have wanted to get a good view of it, Hoagland drew a horizontal line 90 degrees from the structure's vertical axis. It led him to the center of four small regular mounds in the pattern of a cross, housing a fainter central mound—this itself seemingly at the center of a group of ten geometrical pyramidal forms. He christened this collection of features "the City" and described it as a

> remarkably rectilinear arrangement of massive structures, interspersed with several smaller "pyramids" (some at exact right angles to the larger structures) and even smaller conical-shaped "buildings." The entire gathering measured something like four by eight kilometers—a strikingly rectangular pattern created by numerous features at right angles to each other, including aligned corners and even "streets" running roughly north and south.[2]

The easternmost structure of this grouping was termed by Hoagland "the Fort." It is a straight-edged feature that seems to consist of two huge walls, each roughly a mile in length, meeting at the southwest corner, enclosing a regular inner space, like the keep of a giant castle.

More discoveries were to follow.

LINES ON THE LANDSCAPE

Hoagland's next find was the so-called Cliff, 14 miles east of the Face—that is, on the side opposite the city. He noticed that this curious formation lies strangely untouched by, and at right angles to, a splash of crater ejecta material—suggesting that it was built after the crater was created.

The Cliff, which lies on an axis parallel to the Face, might be a thin wedge-shaped mesa or a gigantic wall. It seems to act as a backdrop to the profile of the Face as seen from the City, along a line that runs from the "City Square," through the mouth of the Face, and then on to the center of the Cliff.

Hoagland used computer technology to re-create the Martian sky to see if this horizontal line could have any astronomical significance. He calculated that a viewer positioned at the city center would have seen the sun rising out of the Face's mouth at dawn on the summer solstice approximately 330,000 years ago.

ENTRY TO THE CITY

The main structures of the City are found in a circle around the "City Square," as Hoagland terms the cross-shaped pattern of small mounds.

The surrounding large structures, each roughly the same size as the Face, are straight-sided and appear to be pyramidal in form. The only exceptions are a feature opposite the Face—which is oval, like the Face itself—and the Fort, which resembles a huge triangle, with what looks to be two sides of immense walls enclosing an inner space, the third side being more built up and irregular.

Dotted around the feet of the monstrous pyramids that define the City are sixteen small oval mounds. They are set in no immediately obvious pattern save for the City Square, with its four mounds in a cross arrangement. These mounds are so small that no detail can be gleaned from them other than their position and size. And yet, as we shall later discuss, they are of prime importance in the AOC debate.

On first view the City is not overly eye-catching. Closer inspection, however, brings a surprising number of features to the fore—features that sometimes seem to click into a semblance of order.

The Fort, again, is particularly noteworthy. Its two gargantuan walls are perfectly straight, and the inscribed hollow they house is parallel with the outer walls and regular in shape. Wind may be able to scour the outer parts of a rock formation in all manner of ways, but what geological force could excavate the *interior* of such a formation into such exact conformity with its exterior?

THE HONEYCOMB

The part of the Fort that looks most "artificial" is its western side. It was here, perusing DiPietro and Molenaar's reprocessed *Viking* images in 1983, that Hoagland discovered what he terms "the Honeycomb." This peculiar formation looks like a series of cubical cells arrayed in a deliberate architectural configuration against the Fort. It has been disputed by other AOC researchers who argue that it is merely an anomaly of the SPIT processing program.

The *McDaniel Report* provides a balanced view: "Carlotto's photoclinometric and computer enhancement results do not reveal the cell-like structure seen in the SPIT-processed images. They do, however, reveal a series of regular, terrace-like bands at the southwest corner of the Fort in the area associated with the 'honeycomb.' This may be part of the fine detail that generated the honeycomb effect, or it may be an independently existing, but equally anomalous, feature."[3]

McDaniel and a colleague, Dr. Horace Crater, did some research of their own in the City area and discovered a number of additional characteristics smacking of artificiality—for example, specific measurements between the various small oval mounds housed around the complex, and meaningful measurements in the main structures. We will consider these measurements in more detail in a later chapter.

NO EXPLANATION

What are the chances of such artificial-looking objects occurring naturally—particularly when there are so many of them in such close proximity to one another? Since NASA's official view is that *all* the structures are 100 percent natural, its scientists have struggled to find natural solutions to this problem. Cal Tech's Dr. Arden Albee sums up:

> Cydonia—the "structures"—this pattern that's there, was looked at way back early in the *Viking* days as an area in which a strange kind of erosion had occurred, and had not been fully understood. So from a geological point of view, the area is one which is of scientific interest and would have been photographed Face or no Face. It does indeed have some strange structures, but they appear to be the effect of some kind of erosion—whether it's wind erosion or exactly what isn't very clear. The people who have looked at these Cydonia "structures" are looking at them as erosion features, trying to understand.[4]

So, officially, as of yet there is no natural geological explanation for the Cydonia structures. All that NASA can really offer to oppose the well-thought-out and thoroughly argued case that has been made by scientists like Carlotto and DiPietro is an *assumption* that a natural explanation will eventually be forthcoming. Maybe so. But it is also possible that other information may leak out about the Face that will take it out of the realm of the natural forever.

12

The Philosophers' Stone

All is number.

PYTHAGORAS

At that time shall the stones speak . . . the secrets of the deep shall be revealed.

MERLIN, IN GEOFFREY OF MONMOUTH'S
THE HISTORY OF THE KINGS OF BRITAIN

Hic lapis exilis extat precio quoque vilis
Spernitur a stultis, amatur plus ab edoctis.

Here stands the stone from heaven,
'Tis very cheap in price!
The more it is despised by fools,
The more loved by the wise.

ARNALDUS DE VILLANOVA, ALCHEMIST, D. 1313

CARL Sagan was a dedicated opponent of all those who suggested that the monuments of Cydonia could be evidence of intelligent extraterrestrial life. Yet in several of his works of fiction and nonfiction, Sagan argued for the likely existence of intelligent life elsewhere in the universe. *Contact,* released as a feature film after his death in 1997, describes the first encounter—in the form of a binary code received by radio-telescope—

between mankind and an alien civilization. This is, in reality, how most scientists today predict we will ultimately make contact with an alien intelligence.

In *Cosmos,* his best-known work, Sagan states:

> There is something irresistible about the discovery of even a token, perhaps a complex inscription, but, best by far, a key to the understanding of an alien and exotic civilization. It is an appeal we humans have felt before.[1]

Sagan then refers to the discovery of the Rosetta Stone in 1799 by a French soldier working in the Nile Delta at Rashid (Rosetta). On this stela the same inscription appears in three languages—ancient Egyptian hieroglyphs, demotic (the ancient Egyptian cursive script), and Greek. It was this stone that enabled the French scholar Jean François Champollion to crack the code of the hieroglyphs and translate them for the first time. Sagan continues:

> What a joy it must be to open this one-way communication channel with other civilizations, to permit a culture that had been mute for millennia to speak of its history, magic, medicine, religion, politics, and philosophy.
>
> Today we are again seeking messages from an ancient and exotic civilization, this time hidden from us not only in time but also in space.
>
> If we should receive a radio message from an extraterrestrial civilization, how could it possibly be understood? Extraterrestrial intelligence will be elegant, complex, internally consistent, and utterly alien.
>
> Extraterrestrials would, of course, wish to make a message sent to us as comprehensible as possible. But how could they? Is there in any sense an interstellar Rosetta Stone?
>
> We believe there is. We believe there is a common language that any technical civilization, no matter how different, must have. That common language is science and mathematics. The patterns of nature are everywhere the same.[2]

Sagan is writing about receiving an alien message, expressed in the universal code of mathematics, in the form of a radio signal. Yet what if the message was not sent as a radio signal but was built into the surface of a neighboring planet?

CULTURAL BLINDNESS

Could it be possible that we are so educated to expect communication via a radio-telescope that when we get other signals we ignore them?

Is a humanoid face on Mars so obvious that it is passed over without thought? For scientists waiting for a series of regular beeps to surface from an oceanic roar of electronic background noise, would the Cydonia landscape be just too clear a signal—so clear that it seems ridiculous?

In his book *Lila,* author and philosopher Robert Pirsig tells of sailing into port at Cleveland, when because of misreading the chart he believed he was actually some 20 miles upshore in a completely different harbor. Yet the landscape seemed to tie in with the chart—until he remembered having discounted discrepancies between the map and the land, convincing himself that changes had been made to the shoreline since the chart was produced.

How could he have made such a mistake in the daylight? Didn't he have his eyes open? Writing about himself in the third person Pirsig states:

> It was a parable for students of scientific objectivity. Wherever the chart disagreed with his observations he rejected the observation and followed the chart. Because of what his mind thought it knew, it had built up a static filter, an immune system, that was shutting out all information that did not fit. Seeing is not believing. Believing is seeing.
>
> If this were just an individual phenomenon it would not be so serious. But it is a huge cultural phenomenon too and it is very serious. We build up whole cultural intellectual patterns based on past "facts" which are extremely selective. When a new fact comes in that does not fit the pattern we don't throw out the pattern. We throw out the fact. A contradictory fact has to keep hammering and hammering, sometimes for centuries, before maybe one or two people will see it. And these one or two have to start hammering on others for a long time before they see it too.[3]

Are our scientists so bound to existing beliefs that they are immune to the facts being unearthed at Cydonia? Because they were expecting a radio signal, and because it was the preconception of the time that there was never life on Mars, did figures such as Sagan simply filter out what they were seeing when possible artificial structures were first identified on the Red Planet? The *McDaniel Report* asks us to consider what would have happened if the same information had come in from much farther away in a more "conventional" form:

> Imagine that a digital pattern of radio signals originating in deep space has been received via the SETI radio telescopes. Translated into images by computer, the first image is that of a humanoid face wearing a peculiar headpiece, and the second is a pentagonal diagram [like the D&M Pyramid] having unique proportions and redundant mathematical constants. . . . Would NASA file these images away, like some lost Ark, claiming they were merely "a trick of radiation and noise?" And if a portion of the signal appeared to have been distorted by interstellar static, would NASA stop listening on that frequency, saying the message was not complete enough?

THE LANGUAGE OF STONE

Where are ancient Egypt's radio transmitters? Quite simply, the knowledge we have of ancient Egypt was not received by radio. Instead we have relied on the survival of artifacts bearing inscriptions and other useful data. But even if no hieroglyphs had survived at all, we still would have been able to learn a great deal about the Egyptians from their colossal buildings. A stone pyramid, in other words, may not be able to travel through interstellar space, but as a "signal" of intelligence it lasts longer than a radio transmission—being one of the most stable forms in nature. If any race, human or alien, wished to leave a message in stone, they could choose no better vehicle than a pyramid to transport it down through the ages.

There is, of course, the possibility that any artificial structure will contain cultural references and "messages" even if these are unintentional. For example, anyone decoding a structure such as the Parthenon in Athens would be able to derive from its construction the fact that it was built by an intelligent culture with knowledge of mathematics and geometry. As Sagan is the first to admit: "Intelligent life on Earth first reveals itself through the geometrical regularity of its constructions."[4]

KEYSTONE

In 1988, Erol Torun, a cartographer and systems analyst for the U.S. Defense Mapping Agency, read Richard Hoagland's book *The Monuments of Mars.* Later he wrote to Hoagland, saying:

> While I was impressed with most of the images presented and your description of them, the object that especially caught my attention was the D&M Pyramid. I have a good background in geomorphology and know of no mechanism to explain its formation.[5]

The appearance of the 1.6-mile-long D&M pyramid on frame 70A13 is indeed puzzling. It has been calculated that it incorporates more than a cubic mile of material and that its apex towers almost half a mile above the surface of the surrounding plains. It is strangely buttressed at the base of each of its five corners, adding to its architectural majesty.

Its most fascinating feature is to be seen on the southwestern facade forming the base of the pentagonal structure—the tip of which points toward the Face. This shows quite clearly a regular triangular plane that is similar to the side of a terrestrial pyramid. Quite frankly, from this angle, it looks artificial—no doubt about it. However, as with the Face, the evidence from the rest of the structure is not as clear. Damage to its eastern, shadowed, side spoils its regularity—and the fact that DiPietro and Molenaar first thought the pyramid had only four sides shows how indistinct this area is. It is also penetrated by a deep hole, previously thought to be a crater. Carlotto's photoclinometric reconstructions have raised the extraordinary possibility that this hole could in fact be a tunnel. Subsequently there has been speculation that the pyramid might originally have been a *hollow* structure that partially collapsed at some point in its history—the collapse causing its obvious deformity and the apparent shortening of its right "leg" (the missing portion presumably being hidden under dust and debris).

Such ideas cannot be more than speculation until higher-resolution pictures are obtained. What is not in doubt, however, is that the pyramid does have an unmistakably pentagonal outline. It was this shape, above all the others at Cydonia, that attracted Torun's attention.

WEIRD GEOLOGY AGAIN?

Torun began his analysis by systematically researching known geological processes to see if any could have formed a pentagonal, five-sided pyramid. To this end he examined the effects of five different factors: water, wind, mass wasting (natural slippage of material due to faults, etc.), volcanism, and even crystal growth. His results were conclusive:

> Fluvial [river water] processes can be ruled out as mechanisms for forming the D&M Pyramid as there are no indications that water ever flowed one kilometer deep in Cydonia Mensae (one kilometer being the approximate height of the D&M Pyramid). It is also true that sharp-edged multifaceted symmetrical shapes are not characteristic of fluvial land forms.

The D&M Pyramid is located on what has been described as "knobby terrain," which stood above the once-flooded Cydonian plain. Though this area does show signs of water erosion (due to coastal tides), it is very slight.

As for wind erosion, a favorite explanation of many scientists, Torun concluded:

> No dune will ever form a symmetrical polyhedron resembling the one under study. Flat sides and straight edges are unobserved in terrestrial or Martian sand dunes.
>
> Prevailing winds are not likely to have shifted periodically with perfect symmetry and timing. Even if this seemingly impossible condition were satisfied, another factor would prevent such an object from forming. . . . Locally reversed airflow can cut a flat surface perpendicular to the wind direction on the leeward side of a wind-cut hill. This locally reversed airflow, and associated surface-level turbulence, would prevent the formation of this hypothetical five-sided ventifact. Each time the wind shifted to a new direction, the reversed airflow would start erasing

> the edges formed by other wind directions. The end result would not be a pyramidal hill, but rather a round one.[6]

Torun's conclusions on this matter correspond to NASA's own inability to reproduce pyramidal landforms in a wind tunnel. Similarly, no features formed due to "mass wasting" could account for a five-sided structure—the likelihood of five geological faults all causing land to slip to produce a bisymmetric polygon are next to impossible.

Finally, as for "volcanism" and "crystal growth," there is simply no evidence of volcanic activity in Cydonia, just as there are no naturally occurring pentagonal crystals. Even if there were, crystals are regular; the D&M, on the other hand, although bisymmetrical, contains different side lengths and angles.

What about *unknown* erosional forces? After all, Mars and Earth are two different planets. Torun replies:

> All observations to date of the geophysics of Mars, its gravity, meteorology, geomorphology, etc., indicate that Mars is a place where the laws of physics and principles of geomorphology as we understand them apply, with minor variations due to gravity and atmospheric density and content. It is illogical to assume that there is one small place on the surface of Mars where these same principles are being violated.[7]

ALIEN ARCHITECTURE

Not content to let the matter rest there, Torun tested the supposed artificiality of the D&M Pyramid even further with a series of revealing questions:

1. Is the object's geometry inconsistent with known landforms and geomorphological processes?
2. Is the object aligned with the cardinal directions and/or with significant astronomical events?
3. Is the object co-located with other objects that are also inconsistent with the surrounding geology? And if so, are they geometrically aligned with one another?

4. Does the object's geometry express mathematically significant numbers, and/or the symmetries associated with architecture?

The first question is easily answered. As we have seen, no known geomorphological processes account for the pentagonal form of the D&M Pyramid. In answer to the second question, the Pyramid is indeed aligned to the Martian cardinal directions. As for question three, Torun states:

> The front of the D&M Pyramid has three edges, spaced 60 degrees apart. The center axis points to the Face. The edge on the left of this axis points toward the center of a feature that has been nicknamed the "City" by the Cydonia investigators. The edge on the right of the center axis points toward the apex of a dome-like structure known as the "Tholus."[8]

In Torun's view these three alignments are remarkable evidence of artificiality. After all, how many random geological features could fit together and point at one another so snugly? Surely it would be rare to find an anomalous structure, inexplicably unique geologically, meaningfully aligned to the cardinal directions and to other unique structures in the vicinity, that nevertheless turned out to be 100 percent natural?

Rare, one might say, but not impossible.

But what if this structure also meets the criteria in question 4?

RECONSTRUCTIONS

To answer this last point, Torun had to model the original shape of the damaged and eroded pyramid—arguing, correctly, that this is now standard practice in reconstructive archaeology, especially in sites connected to astronomical alignments or specific geology. Once the model was created he measured it to establish whether or not it possessed any significant mathematical features. He was wary of delving into complicated "numerology" and confined himself to the following basic measurements only:

1. The values of observable angles expressed in radian measure.
2. Examining the ratios formed between the observable angles for equality with mathematically significant numbers.

3. Examining the sine, cosine, and tangent of measured angles for the presence of mathematically significant numbers.

"These approaches," explains Torun, "were selected due to their simplicity, their validity in number bases other than decimal, and their independence from our convention of expressing angles as a portion of a 360-degree circle."

Taking an orthographic projection of the pyramid, Torun measured all visible angles (with a calculated error of ± 0.2 deg).[9] A variety of angles offer a variety of ratios. On the premise that an artificial monument would express meaningful measurements and proportions, Torun began to look into these ratios.

To understand his results, it is first necessary to make a brief excursion into the realms of sacred geometry.

SACRED NUMBERS

In the fifth century B.C., initiates of the mathematical and geometric mysteries of the philosopher Pythagoras communicated their fellowship with a secret sign. On meeting a stranger a Pythagorean would offer him an apple. If the stranger was also a Pythagorean he would cut the apple laterally across its core to reveal the pips laid out in the shape of a pentagram.[10]

The pentagram was a sacred symbol of the Pythagoreans, as it contained within it references to the mathematical measurement known as the "golden section," or *phi* ratio:

> There seems to be no doubt that Greek architects and sculptors incorporated this ratio in their artifacts. Phidias, a famous Greek sculptor, made use of it. The proportions of the Parthenon illustrate the point.[11]

Indeed, it was after Phidias that *phi* was named. *Phi* has to do with proportion—being the ideal ratio between two lengths that produces the greatest aesthetic effect on the eye when incorporated into the measurement of a work of art or architecture. A rectangle made of sides whose relationship to one another is based on the *phi* ratio will be more visually pleasing than any other rectangle.

Look at line ABC:

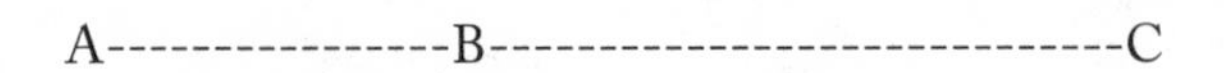

The *phi* ratio is demonstrated in a figure in which the length AB has the same relationship to the length BC as the length BC has to the entire length AC. For this to be so, the ratio has to be precisely 1:1.61803398.

Why *phi* produces such an aesthetic effect is a mystery, but the Pythagoreans saw it as reflecting the harmonies of nature—the same figure is found widely throughout the natural world in organic life. The spiraling of a snail's shell incorporates *phi,* as do the distances between leaves on branches.[12] The proportions of the human body also relate to *phi*—which, for example, is the ratio of the length of the body from the head to the navel and from the navel to the feet.

Thus the Pythagoreans claimed "all is number" and used geometry as a metaphor for higher concepts and metaphysical assertions. To them *phi* expressed beauty—not as a subjective opinion as in "beauty is in the eye of the beholder," but as a quality intrinsic to the object itself. Beauty is in the beheld.

VESICA PISCIS

Phi is also generated by the most widely used and most sacred of geometric forms—the *vesica piscis,* "vessel of the fish"—consisting of two overlapping equal circles, the centers of which each stand on the circumference of the other circle.

To the ancient geometers this device represented the union of spirit and matter, heaven and earth.[13] In it were generated not only *phi,* but the constants of the sacred square root series of 2, 3, and 5 and the five regular solids.[14] This sacred figure was used as the basis of various ancient monuments including the St. Mary Chapel at Glastonbury Abbey and, according to John Michell, an expert in sacred proportion, the Great Pyramid at Giza.[15]

The Pythagorean secret sign, the cutting of the apple, was the transmission of a shared wisdom—that of the knowledge of the numerical harmonies of nature revealed through the *phi* ratios of the pentagram and, by extension, the *vesica piscis*. This message was nonverbal. All you needed to grasp it was the knowledge of mathematics, the universal language.

But what has this to do with Torun's model of the D&M Pyramid? He claims it has everything to do with it.

ROSETTA STONE

When DiPietro and Molenaar discovered the pentagonal pyramid they noted its dimensions as 1 mile by 1.6 miles.[16] These figures are, of course, extremely close to the golden section ratio.[17] In Richard Hoagland's opinion they may also have a deeper significance. Staring at the "exquisite five-sided bisymmetry" of the D&M Pyramid, he reports:

> Another striking aspect of this "magic" ratio suddenly appeared before me: Leonardo da Vinci's application of these ancient "sacred" proportions . . . to the human form. And suddenly I comprehended an extraordinary possibility: If I superimposed da Vinci's famous figure—"a man in a circle"—over the stark geometric outlines of the D&M, the two conformed. The D&M seems to be a striking geometric statement of humanoid proportions arrayed on an alien landscape almost in the shadow of the central "humanoid" resemblance [the Face].[18]

It was this assertion of Hoagland's that first caught Torun's attention. What was a universal constant of aesthetic proportion doing on an inorganic mountain on Mars? Torun's own findings were to be even more surprising, as the authoritative *McDaniel Report* confirms:

> What Torun discovered was a mathematically rich figure whose geometry contains the mathematical bases for the hexagon, the pentagon, and the classic geometric proportions of the Golden Ratio. Twenty of the model's internal angles, angle ratios, and trigonometric functions redundantly express three square root values, *sqrt* 2, *sqrt* 3, *sqrt* 5, and two mathematical constants, *pi* (the ratio of the circumference of a circle to its diameter) and *e* (the base of the natural logarithms). . . . Except for *sqrt* 2 and *sqrt* 3, the constants do not appear alone, but in seven different mathematical combinations. The most redundant values discovered

> were *e/pi*, *e/sqrt* 5, and *sqrt* 3. These values were repeated four times each in at least two different modes of measurement.[19]

The D&M Pyramid, in other words, seems to be a veritable textbook of the same numerical forms that were deemed sacred by the Pythagoreans because of their universal harmonic qualities.

VERIFICATION

We must admit that we are impressed by Torun's model, with its amazing ability to yield geometric constants. But wouldn't any pentagonal figure produce the same results?

Keith Morgan, an electronics technician, devised a FORTRAN computer program at Howard University, Washington, D.C., to answer this question. Keeping the two front 60-degree angles, he adjusted the "ridgelines" of the opposite face throughout a range of different angles, generating 680 variations on the pyramidal form. His conclusions confirmed the uniqueness of Torun's model showing it to be the *only* pentagonal form with front angles of 60 degrees that could generate the *vesica piscis* and, simultaneously, the values of *phi, pi, e, sqrt* 2, *sqrt* 3, and *sqrt* 5, and the only one which could represent them all (save *phi)* across the three measurements of angle ratio, radian measure, and trigonometric functions![20]

Clearly Torun has uncovered not only a rich geometric minefield, but a unique one in the bargain, a giant rock containing the Pythagorean constants—a true philosophers' stone.

ALCHEMY

In the ancient art of alchemy, it was the task of the alchemist to find the *lapis exillis*—the philosophers' stone—that turned base metals into gold. This stone was said to have "fallen from heaven," like the meteoric Benben stone of Heliopolis that is spoken of in ancient Egyptian tradition, a pyramidal stone associated with rebirth.

The Benben stone bore arcane knowledge about the nature of the universe—"On the stone is encoded the cipher of life's mysteries"[21]—and it was supposed to redeem spirituality from base matter, the pecuniary aspects of the process being metaphors for spiritual transformation.[22]

Now this pyramidal *lapis,* "the cipher of life's mysteries," is depicted as a stone—and yet it encompasses all matter, being composed of *"de re animali, vegetabili et minerali."*[23] It was also said to grow from "flesh and blood" and to possess a body, soul, and spirit.[24] The *lapis* is thus intrinsically connected with rebirth, new life, and growth.

Strangely, Torun finds similar qualities referred to in the measurement *e/sqrt* 5 found in the Martian pyramidal stone:

> The relationships between *e* and *sqrt* 5 may also be suggestive of biology. Five-sided symmetry is not characteristic of non-living systems. Life-forms on Earth, however, often exhibit five-sided symmetry, especially in the plant kingdom. The constant *e,* the base of the natural logarithms, is also known as the law of organic growth. It is a way of describing growth where the increment of growth is always proportional to the size of the growing quantity, as is often the case in biological systems. Most formulae devised for the study of organic growth, whether for population studies, or predictions of microbial and plant growth, incorporate the number *e* as a factor. The relationship between *e* and *sqrt* 5 might therefore be interpreted as being symbolic of "the exponential growth of life."[25]

Torun supports his interpretation of these numbers as a biological metaphor by pointing to the fact that the D&M Pyramid possesses another characteristic of living things—bilateral symmetry—and "by the alignment of the D&M Pyramid's axis of bilateral symmetry with the one object in Cydonia Mensae that most clearly resembles a living thing: the Face."[26]

MESSAGE

The Pythagorean philosophers saw the *vesica piscis* (whose organic constants and geometric numbers are mirrored in the D&M Pyramid) as a powerful symbol of the joining of heaven and earth, spirit and matter. The pyramidal philosophers' stone served exactly the same function, and yet, in the rhyme of the fourteenth-century alchemist Arnaldus de Villanova quoted at the beginning of this chapter: “The fools rejected it.”

Like the philosophers’ stone, it is Torun’s claim that the D&M Pyramid is some sort of cipher—a latter-day Rosetta Stone—for the whole Cydonia region, revealing a message of intelligent design. As we shall see, the same essential design features recur repeatedly among all the monuments of Cydonia. The structures seem to work together, like the instruments in an orchestra, to create an infinite mathematical symphony.

13

Coincidences

Gentlemen, you do not have a science, unless you can express it in numbers.

ARTHUR EDDINGTON, BRITISH ASTRONOMER WHO VERIFIED EINSTEIN'S GENERAL THEORY OF RELATIVITY

LET us remind ourselves of the mathematical characteristics of the D&M Pyramid. Among other features, its angles and dimensions yield a total of 10 *pi* ratios, 10 *e* values, and 4 *e/pi* values. It also redundantly "prints out" the values of *sqrt* 2, *sqrt* 3, and *sqrt* 5.

Such insistent repetition of geometrically significant data is not a normal characteristic of naturally formed structures. Moreover extremely accurate measurements from the *Viking* photographs indicate another curious indicator of intelligent design: the apex of the D&M Pyramid stands at 40.86 degrees north latitude. The tangent of 40.86 is 0.865—the precise value of the ratio *e/pi* that is repeated four times in the internal structure of the pyramid.[1]

As the Artificial Origins at Cydonia researchers point out, it is almost as though the great pentagonal monument is telling us that "it knows where it is" on Mars.[2]

TIME FOR T

Another notable point about latitude 40.86 degrees north as it runs through the apex of the D&M Pyramid is that it is subtended from the monument's nearest corner-diagonal by an angle of precisely 19.5 degrees. This is an angle that crops up several times elsewhere within the structure.

It is also a highly significant angle within a field of mathematics known as "energetic-synergetic geometry" that was pioneered by the American engineering genius R. Buckminster Fuller (1895–1983). The system takes as its basic unit the tetrahedron (a pyramid shape with four sides including the base—each side being an equilateral triangle) and builds from it a number of astonishing structures, most famously the geodesic dome.

A curious "rule" or constant has been revealed by this geometry and commented on by Richard Hoagland, Stanley McDaniel, Erol Torun, and other AOC researchers. The rule is that when you place a tetrahedron inside an exactly circumscribing rotating sphere so that one of its four vertices touches either the north or the south pole of that sphere, then the other three vertices, each separated by 120 degrees of longitude, will be found at latitude 19.5 degrees south (when the first vertex is at the north pole) or at latitude 19.5 degrees north (when the first vertex is at the south pole).[3] The figure of 19.5 is therefore known as *t*, the tetrahedral constant.[4]

MOUNDS

Torun and Hoagland have always claimed that the tetrahedral numbers yielded by the D&M Pyramid must be significant. This claim, in our view, gains in credibility from recent discoveries by Horace W. Crater, a professor of physics at the Tennessee Space Institute. Working with Stan McDaniel, Crater has found the same specific measurements cropping up in other structures in Cydonia—particularly in the City, with its enigmatic complex of sixteen oval mounds (four of which are directly aligned with the D&M Pyramid).

Hitherto we have only commented in passing on the existence of these bright, uniformly shaped mounds, each 300 to 700 feet in diameter and 100 feet high, dotted around the foothills of the City and stretching out toward the south. Four of them form the regular cross shape of the City Square lining up not only with the D&M Pyramid but also, remarkably, with the mouth of the Face.

MISSED TARGET

When NASA re-imaged sections of the Cydonia landscape in April 1998 (see under heading "Unexpected News," chapter 15), the four mounds forming the cross-hairs of the City Square were selected, on the advice of pro-artificiality scientists, as a rather apt target to follow the controversial re-imaging of the Face.

Unfortunately *Mars Global Surveyor* missed the Square and caught a swathe of land about a kilometer to its left (as seen from above), which included just a single mound and a couple of the least impressive outcrops of the City. Though other intriguing objects dot the surface of this image, unseen by the earlier *Viking* orbiters (such as a strange ring of small pyramidal structures and a larger pyramidal structure on the edge of a rocky outcrop for which we will have to await further analysis) little information was obtained on the enigmatic mounds themselves that could aid classification of these features and their alignments.

The only mound captured by *Mars Global Surveyor* is seen to be a regular, oval-shaped ridged knoll—and, unfortunately, as we have no other high-resolution images to compare it with, it is impossible to tell if it is a natural formation or whether it is similarly structured to the other mounds photographed by *Viking* and thus suggestive of artificiality.

The one thing that the mounds do tell us clearly about themselves, however, is their own precise locations on the surface of Mars. These locations were studied from the original *Viking* frames by Horace Crater and were reported on by Crater and McDaniel in their joint paper "Mound Configurations on the Martian Cydonian Plain: A Geometric and Probabilistic Analysis."

"THEIR ARRANGEMENT WAS NOT NATURAL . . ."

Probably no one is better qualified to evaluate the patterns formed by the mounds than Horace Crater. A specialist in theoretical particle physics, he is a world expert on the transformation of experimental data patterns into mathematical forms, from which further patterns can then be predicted.

"Like many," says Professor Crater, "I was interested in the controversy surrounding the Cydonian Face, but at a distance. It was not until late 1993 that my involvement with the Mars anomaly research began."

Crater started out skeptical, saying of Torun's reconstruction of the D&M Pyramid:

> It was my suspicion that proportions with such redundancy could occur with reasonable odds in any semisymmetrical five-sided figure. Of the various five-sided figures I examined, many showed proportions like

> those of Torun's measurements. As I increased the precision of my calculations, however, I came up with a surprising result. At greater levels of precision only the Torun model appeared with significant redundancy.
>
> This unexpected result stimulated my interest in the Cydonia region. I began to investigate a number of small mound-like features found there. These "mounds" are small enough to make measurements of their geometric relationships relatively precise, within a determinable margin of error. What I found astounded me. Their arrangement was not random.[5]

ANALYSIS

In his paper Crater relates how he began his investigation by labeling the sixteen mounds *A* through *P*, not in any strict order due to their positioning on the planet, but in the order he studied them. His first target was the *E-A-D* group of mounds—those closest to the D&M Pyramid, some miles south of the City. As Hoagland had shown as early as 1992, these three mounds form a perfect isosceles triangle.[6]

Crater based his measurements of *E-A-D* on orthographic prints, which corrected camera tilt to establish a workable Mercator projection, and found that the angles of this triangle were as follows: 70.9 (± 2.9) degrees; 54.3 (± 2.2) degrees; and 53.5 (± 2.2) degrees. These results were strikingly similar, he realized, to the angles of the plane formed inside a tetrahedron when you take its cross section from one axis so that it bisects the opposite face. These angles are, respectively, 70.5 degrees, 54.75 degrees, and 54.75 degrees. Furthermore, when the angles of the ideal tetrahedral cross section are expressed in radians, "We see that all of them are simple linear functions of [the] tetrahedral constant, *t*, equivalent to 19.5 degrees."[7]

Because one isolated result proves nothing, Crater devised a number of tests to see how often a "tetrahedral" triangle could be created randomly, defining a tetrahedral triangle as

> any triangle whose angles in radians are given in simple terms of quarter, half, or whole number multiples of *pi* and *t*.[8]

Crater's tests were thorough and professional (as might be expected of a scholar whose job is the calculation of patterns).[9] He randomly generated 100,000 three-mound placements on a computer, finding just 121 randomly occurring *E-A-D* triangles. Then he analyzed 4,460 actual triangles formed from natural Martian features, of which only two were tetrahedral *E-A-D* triangles. Based on these odds Torun reckoned that the chances of the *E-A-D* triangle occurring naturally was "slightly more than one in 1,000."[10]

This was not an impressive result, and did not rule out the possibility of coincidence. But more was to come.

TETRADS, PENTADS, AND HEXADS

Crater's next step was to introduce mound *G*, which nestles at the feet of the southernmost of the large city structures, thus forming the tetrad *G-A-D-E*. It contains two identical right-angled triangles, *A-E-G* and *G-A-D* and its geometry is entirely determined in terms of *t* and *pi*, as is also the case for the geometric divisions of a tetrahedron.

Crater now included the next closest mound—mound *B*, to the right of triangle *E-A-D*—to form a pentad *G-A-B-D-E*. Like the cogs of some great wheel meshing together, triangles *A-D-B* and *E-A-B* exactly mirror triangles *A-E-G* and *A-G-D*. What's more, all the angles within the pentad also turn out to be functions of *t*.[11] Some wider plan must lie behind this setup, Crater suspects, because:

> The geometry that most optimally describes the mound placements suggests, with stubborn redundancy, the geometry hinted at in Torun's model of the D&M Pyramid.[12]

Next to be analyzed was mound *P*, found to the west of mound *G*. Here, too, the results are confirmatory: triangle *P-G-E* is a mirror of *G-E-A* and *E-A-B*. The odds of such a hexad forming naturally, Crater estimates, are about 200 billion to one.[13] These triangles also repeatedly include the significant angle of 19.5 degrees.[14]

The final development came in February 1995. While studying Crater's results, Stan McDaniel realized that the pattern formed by five of the Cydonia mounds (*G-A-B-D-E*) appears to imply a rectangle, even though two corners of that rectangle are "missing." Using the geometrical analysis performed by Crater, the proportions of the grid were found to be a significant figure in terrestrial sacred architecture—1:1.414, or 1 to the square root of 2.[15] As the reader will recall, *sqrt* 2 is one of the values repeatedly "printed out" by the geometry of the D&M Pyramid.

THE MESSAGE AND THE CONSPIRACY

Following up on Torun and Crater's pioneering work, Richard Hoagland set about combing the Cydonian plain for more alignments that might make sense in terms of tetrahedral geometry.

His first discovery was that the angle between the so-called cliff to the east of the Face and a "tetrahedral pyramid" found on the far lip of the crater on whose ejecta blanket the cliff lies is 19.5 degrees: *t*, the tetrahedral constant.

Hoagland also claims that the "teardrop" on the right side of the Face lies at a point that is exactly equidistant between the City Square and the D&M Pyramid—this distance being 19.5 arc minutes of the circumference of Mars! A second measurement, from the teardrop to the great buttress of the D&M pyramid, corresponds with 1/360th of the polar diameter of Mars.[16]

But this system of dividing up circles and spheres into 360 degrees is surely an Earth-based invention . . . isn't it? Therefore, even if we accept the "way-out" view that the Cydonia monuments are artificial, how can we explain that their presumably alien builders used the same 360-degree system that we do, and even followed geometrical conventions that are of venerable antiquity here on Earth?

Torun and Hoagland came to the conclusion that a message was deliberately being sent, quite possibly targeted at "us," and that the circumference of the planet was continually referred to in relation to the tetrahedral constant for a specific purpose. "All this seems to be directing us," Hoagland theorized in 1987, "to place the inscribed tetrahedron in a planetary sphere such as Mars itself. . . ."[17]

On Independence Day, 4 July 1997, NASA's lander *Pathfinder* touched down in the once catastrophically flooded Martian channel known as Ares Vallis. Richard Hoagland was the first to point out that *Pathfinder* has a pronouncedly tetrahedral design with distinctive solar panels in the form of equilateral triangles. Moreover, its landing site in Ares Vallis is located at 19.5 degrees north latitude.[18]

Probably NASA meant nothing by this. Still, we cannot deny that the act of placing a tetrahedral object on Mars at latitude 19.5 contains all the necessary numbers and symbolism to qualify as a "message received" signal in response to the geometry of Cydonia. Moreover, such a game of mathematics and symbolism is *precisely* what we would expect if NASA were being influenced by the sort of occult conspiracy that Hoagland, for one, is always trying to expose.

PART THREE

Hidden Things

14

Disinformation

The broad mass of a nation . . . will more easily fall victim to a big lie than to a small one.

ADOLF HITLER, *MEIN KAMPF*, 1925

COULD NASA know more about Cydonia than it has admitted? Could it have discovered something there that it has decided to withhold from the public?

In 1938, as Europe readied herself for war, the peoples of the New World found themselves threatened not by some maniacal führer seeking to establish a new order of darkness, but by invaders from Mars. It happened when Orson Welles broadcast his own adaptation of H. G. Wells's *The War of the Worlds* on the radio. The radio-play was so realistically presented that many believed it to be a genuine news report. The widespread panic that ensued revealed what a two-edged sword mass communication could be. It brought people together, but its power to influence vast swathes of the population was clearly immense.

In Germany, Goebbels churned out propaganda films and fed them to the masses, exaggerating resentments and xenophobia (present throughout Europe at this time), and twisting nationalist sentiments to result eventually in the Holocaust. What Hitler had said in 1925 was turning out to be literally true—people were believing the "big lie."

But propaganda was not an invention of World War II and did not end with it. This begs the question of whether NASA scientists today could be abusing their authority—leading the people on or even deliberately lying over Cydonia and other issues? If Welles managed to convince 1930s America that it was being invaded from outer space even though there was no invasion, then it seems obvious that a government should be able to find ways to hide or devalue information concerning contacts with beings from other planets, or traces of intelligent life found on Mars, or that some new fact has been uncovered in explorations of Mars that is of enormous significance for all mankind.

Generally speaking, government agencies find it easier and preferable to reinforce already held beliefs than to introduce new ones. We therefore have no difficulty envisaging situations in which NASA might decide *not* to share everything it knows with the public—for example, if it believed that a specific piece of information might be socially, or politically, or economically destabilizing. We can also imagine other less honorable motives that might lead officials to hide the truth about certain types of discovery.

Because such things are possible, and because discoveries have been hidden and hushed up in the past, we think it would be naive to place any great confidence in NASA's repeated assurances that the monuments of Cydonia are all natural landforms. Like other big state bureaucracies, NASA has lied and will lie again. We think the evidence suggests that it has lied about Cydonia ever since the Face on Mars was first discovered.

DUTY TO WITHHOLD

NASA is not some *Starship Enterprise* on a "mission to seek out new worlds and civilizations, to boldly go where no man has gone before." On the contrary, NASA is the disturbed child of two dysfunctional parents—paranoia and war.

NASA was formed in 1958 at the height of the Cold War when all advances in space science were spin-offs from the development of more efficient killing machines. The exploration of space itself was directly linked to defense policy.

To a certain extent, this Cold War mentality still prevails. Thus, although it is funded from public taxes, NASA is finally not responsible to the people but to the government of the United States. Nor does any law compel it to share information openly with the public. In Section 102 (c) (a) of the Act of 29 July 1958 (The Space Act), which formed NASA, we read:

> NASA is charged with the making available to agencies directly concerned with national defense of discoveries that have military value or significance. . . .
>
> Information obtained or developed by the Administrator in the performance of his functions under this act shall be made available for public inspection except:
>
> a) information authorized or required by Federal statute to be withheld, and
>
> b) information classified to protect the national security.

So it seems that NASA actually has a "duty to withhold" certain categories of information.

THE BROOKINGS REPORT

NASA scientists cannot know for sure, on present evidence, whether or not the structures of Cydonia are natural or artificial. Many intelligent people therefore suspect there must be some very strong reason why NASA has for so long failed to test the AOC hypothesis.

It has been suggested that a 1960 Brookings Institute report may contain a possible clue. The report is entitled *Proposed Studies on the Implications of Peaceful Space Activities for Human Affairs.* Amid other advice it urges that if NASA should ever discover evidence of extraterrestrial life, it should seek to control this information for reasons of public security, considering the plight of "societies sure of their place in the universe, which have disintegrated when they had to associate with previously unfamiliar societies espousing different ideas and different life ways."[1]

At the level of policy and strategy the Brookings report recommends that NASA should always ask, and consider very carefully

> how such information, under what circumstances, might be presented to or withheld from the public for what ends. What might be the role of the discovering scientists and other decision-makers regarding the release of the fact of discovery?[2]

The report was commissioned by NASA in 1958 (the year of its inception) from the Brookings Institute in Washington, D.C., and was delivered to the chairman of NASA's Committee on Long-Range Studies in 1960.[3] It includes a subsection starting on page 216 titled "Implications of a Discovery of Extraterrestrial Life":[4]

> Cosmologists and astronomers think it very likely that there is intelligent life in many other solar systems. . . . *Artifacts left at some point in time by these life-forms might possibly be discovered through our future space activities on the Moon, Mars, or Venus.*[5]

The Brookings Report evisages that hard evidence of intelligent extraterrestrial life might have severe effects on political leaderships—shaking society up and causing the public to question entrenched elites:

> The degree of political or social repercussions would probably depend on leadership's interpretation of (1) its own role; (2) threats to that role; and (3) national and personal opportunities to take advantage of the disruption or reinforcement of the attitudes and values of others.[6]

UFO

The policy of secrecy regarding possible alien artifacts stems back some years before NASA was formed. The recommendations of the Brookings Report only echo earlier statements made by the American government.

The "Report of the Meetings of the Scientific Advisory Panel on Unidentified Flying Objects Convened by Scientific Intelligence, CIA, January 14–18, 1953" concludes:

> The continued emphasis on the reporting of these phenomena [UFO encounters] does, in these perilous times, result in a threat to the orderly functioning of the protective organs of the body politic.[7]

Many conspiracy theorists in the United States passionately believe that such conclusions were first drawn six years earlier—in 1947, to be precise.

THE CRASH OF '47

The modern UFO phenomena can be said to have begun with the sighting of nine "saucer-shaped" objects flying over Mount Rainier, Washington, by pilot Kenneth Arnold on 24 June 1947.[8] Two weeks later, rumors began to circulate concerning an alien spaceship that had supposedly crash-landed in Roswell, New Mexico.

The "Roswell Incident" has been given much public attention recently due to the celebration of the fiftieth anniversary of the crash in 1997. It is an understatement to say that it has caught the imagination of the present generation: an increasing variety of claims about the crash have been put forward in recent times, most of which accuse the U.S. government of covering up the evidence. It was to refute such claims that the Pentagon embarked on a four-year research program to dismiss these theories.

In a report titled *Roswell: Case Closed,* published on 24 June 1997 (fifty years to the day after Arnold's first sighting of "flying saucers"), the Pentagon claims that what crashed at Roswell was a high-altitude weather balloon and that the "alien bodies" reported to have been found beside it were "life-size dummies from top-secret simulated parachute drops."[9]

The crash was discovered by Mac Brazel, a rancher checking for storm damage near the Roswell Army Air Force Base (RAAF). The wreckage that he found consisted of a strange shiny material that was immutable, returning to its original shape when crumpled into a ball. Unable to identify this substance, he handed it in at the air base. On 8 July 1947 the base issued an official army press release stating that a "flying disk" had been found, the local paper's headline stating RAAF CAPTURES FLYING SAUCER ON RANCH IN ROSWELL REGION.[10] Within hours the Pentagon contacted the head of the local radio station and told him to stop broadcasting the news, and a new press release was issued stating that what had really been found was a weather balloon.

A major challenge to this story was mounted by several Roswell locals who vociferously claimed to have seen not just wreckage but also the occupants of the wrecked craft. Frank Kauffman, a civilian working at RAAF at the time, reports seeing the bodies of five aliens being placed into body bags by the military. Also among the witnesses was Colonel Philip Corso (now retired), who was on General MacArthur's intelligence staff during the Korean War and on President Eisenhower's national security staff for four years. He claims to have seen at least one short, gray, hairless alien body after it had been removed from the site and stored at Fort Riley, Kansas:

> At first I thought it was a dead child they were shipping somewhere, but this was no child. It was 4 ft., human-shaped figure with arms, bizarre-

> looking four-fingered hands—I didn't see a thumb—thin legs and feet, and an oversized . . . lightbulb-shaped head.[11]

THE DUMMIES

The Pentagon's counterclaim that the bodies were just "life-size dummies from parachute drops" is an admission that there was at least *something* at Roswell that could be mistaken for alien bodies. But how likely is it that such dummies would have landed right next to a crashed balloon? What were the military doing testing parachutes on the night of a violent storm?[12] If the eyewitnesses can be trusted, why place the dummies in body bags? Moreover, what is to be made of statements from several of the witnesses that one of the "aliens" survived the crash and was seen moving?

The army press officer who issued the 8 July press release in 1947 would later sum up the the many absurdities of the Pentagon's position:

> It's just another cover-up. Any dummy knows what a dummy looks like, and those weren't dummies.[13]

UFO RELIGIOUS CRISIS?

But why would NASA want to cover up evidence of intelligent aliens?

To be sure, the Brookings report does suggest a possible motive. However, the public of the year 2000 does not have the same fears as the public of 1960—and NASA must know this. Surveys in the 1990s suggest that 65 percent of all Americans believe that a UFO did crash at Roswell.[14] In addition, surprisingly large numbers of people, probably running into tens of millions, believe that they have either seen or been abducted by alien entities.

As there is clearly no widespread panic about these matters, how likely is it that there would be panic over the as yet hypothetical discovery of alien artifacts on Mars?

The surveys suggest there would be no panic. On the contrary, such news would probably be received positively even by so-called fundamentalist groups. One particularly instructive report is the *Alexander UFO Religious Crisis Survey: The Impact of UFOs and Their Occupants on Religion.*

Written by Victoria Alexander for the Bigelow Foundation, Las Vegas, Nevada, the report considers responses to questions by 230 leaders of religious communities across America (134 from Protestant churches, 86 from Roman Catholic churches, and 10 from Jewish synagogues). While the relatively small size of this survey means that it cannot be taken as definitive, its results are surprisingly clear. As Alexander sums up:

> The numbers are not just statistically significant; they demonstrate unmistakable trends. Even though this was a pilot study, for the first time there are data concerning the perceived relationship between religion and the existence of intelligent extraterrestrial life. The data are counter to the widely held belief frequently posited by many in the UFO community predicting doom and destruction in the wake of verifiable contact.[15]

A typical Alexander multiple-choice question begins with a proposal and asks respondents to categorize their reactions to it. For example:

> Official confirmation of the discovery of an advanced technologically superior extraterrestrial civilization would have severe negative effects on the country's moral, social, and religious foundations.
> a) strongly agree
> b) agree
> c) neither agree nor disagree
> d) disagree
> e) strongly disagree.

It is notable that 77 percent of the respondents either disagreed or strongly disagreed with this particular proposal. Their answers to 10 other questions convey the same mood:

> The results conclusively demonstrate that the religious leaders surveyed believe that the faith of their parishioners is both sufficiently strong and flexible to accommodate this information. Contrary to the belief widely held in the UFO community, it is highly unlikely that such news would yield a religious crisis.[16]

Some conspiracy theorists believe that the public's changed attitudes are themselves engineered by the authorities through information management. The suggestion is that we are all the victims of a billiant propaganda campaign designed to acclimatize us, slowly, to the reality of intelligent extraterrestrial life. The notion is probably fanciful. Nevertheless, we cannot deny that movies like *Independence Day, Stargate,* and *Close Encounters of the Third Kind,* TV programs such as *The X-Files* or *Dark Skies,* and NASA's decision to release information about possible "primitive" life in Martian meteorites have all contributed to the present relatively open-minded state of public opinion concerning ET contacts.

PROPAGANDA WAR

Our own impression is that NASA has attempted to manipulate public perceptions concerning the issue of artificial origins at Cydonia and that it does seem to be covering something up. We cannot say what it is covering up—perhaps only its own bungles—but the agency appears to have acted dishonestly from the beginning.

The lies began on 25 July 1976 when the first *Viking* photograph of the Face, frame 35A72, was released to the press. As the reader will recall, NASA claimed at the press conference that there was a second photograph, at a different sun angle, proving the Face to be just a trick of light and shadow. More than seventeen years passed before officials finally admitted that such a disconfirming photograph does not exist.

We then see the misfiling of images, so that a *confirming* photograph—frame 70A13—was not in the correct file. This threw researchers off the trail for several years. They also had to deal with certain forms of censorship, as Stan McDaniel recounts:

> The first paper on the subject [of artificial origins at Cydonia], authored by a group called the Independent Mars Investigation Team, reporting for the most part the work done by Vincent DiPietro and Gregory Molenaar, was inexplicably expunged from the published papers of the first Case for Mars Conference in 1984. Subsequent attempts to publish papers on the topic, by scientists with impeccable credentials and a long

> list of published scientific papers, were uniformly refused consideration by the primary American journals of planetary science. These scientists were forced by this censorship to turn to publishing their work in books for the general public, whereupon NASA characterized them as seeking personal gain and running "cottage industries."
>
> Over the course of time, as individual citizens, having read such publications, began to ask questions of NASA, a long string of spurious arguments were put forward against the idea that the Face on Mars might be artificial. The services of that powerful propagandist, Carl Sagan, were evidently engaged in this task. Sagan went about writing and talking about psychological aberrations that make people see faces everywhere, whipping out a deformed eggplant at lectures and claiming it looked like Richard Nixon, thereby proving that the Face on Mars was natural. An amazing scientific feat.
>
> Then, in 1985, Sagan published an article in *Parade* magazine debunking the Face, characterizing anyone who took it seriously as a kind of a "zealot," and including a doctored version of one of the *Viking* frames that used false color to make it look as though the Face is actually not there.[17]

If NASA is so sure that the Face is merely an illusion or aberration of nature, then why resort to blatant fraud in order to convince the public of this? The doctoring of frame 70A13 in the *Parade* article—by overlaying the image with a color filter to obscure details that corroborate frame 35A72—is a particularly unscientific and indeed barbaric act. One cannot even defend Sagan by saying that this frame was supplied to him already doctored by NASA, for Richard Hoagland had personally shown Sagan the original frame prior to the publication of the *Parade* article.[18] Sagan was well aware that 70A13 confirmed 35A72 and had earlier told Hoagland that he found this intriguing.[19]

So why did Sagan lie?

Whatever his motives, he appears later in life to have regretted his actions. In his last book, *The Demon-Haunted World* (1996) he actually praised the Cydonia researchers and said that the Face deserved a closer look.[20] Was he here voicing a personal truth, now unrestricted by the laws of NASA?

THE IMPORTANT MAN

Sagan's role as chief scientific critic of the AOC hypothesis has been inherited by Dr. Michael Malin, head of Malin Space Science Systems. Malin, the private contractor who supplied and operated the camera systems for the failed *Mars Observer* mission (1992–1993), is also the supplier and operator of the camera systems onboard *Mars Global Surveyor.* Dr. Malin has published an image of the Face on his World Wide Web page claiming to show "how the face got its teeth." This is supposed to be a jeering dismissal of teeth-like features identified by Mark Carlotto.[21] Yet instead of addressing those features, Malin singles out what McDaniel describes as "deliberately induced pixel errors."[22] By such tactics the suggestion is conveyed that the idea of the Face having something like teeth derives from "amateurs using extremely poor image enhancement and publishing their defective results in American tabloid magazines."[23]

As we will see in the next chapter, Dr. Malin is the most important man in the world where Mars is concerned. He alone decides where the cameras of *Mars Global Surveyor* will point. And he enjoys another amazing privilege: the right to an exclusive six-month preview of *Surveyor's* images *before* they are shown to the public.

If there is not a conspiracy, then how can it be good for one man to have so much power? How can it be good for one man to be given such a monopoly over knowledge that he becomes the sole amanuensis for the story of Mars?

On a matter of such seminal importance surely we should be hearing other voices.

15

Camera Obscura

SWINDON: What will history say?
BURGOYNE: History, Sir, will tell lies as usual.

GEORGE BERNARD SHAW, *THE DEVIL'S DISCIPLE*, ACT 3 (1901)

IN the early 1900s, in the English village of Cottingly near Bradford, Elsie Wright and Frances Griffith took photographs of fairies at the bottom of their garden. Even great intellectuals such as Sir Arthur Conan Doyle, the creator of Sherlock Holmes, fell for this hoax, which the aging Elsie and Frances revealed the photos to be some sixty or so years later.[1] They got away with it because photography was in its infancy at the beginning of the twentieth century and people lacked the skill to spot an obviously doctored image.

Things changed and people today are very aware of the fact that cameras, especially when linked to computers, can lie and do lie. Hollywood special effects teams such as Industrial Light and Magic prove to us again and again that the impossible can easily be made possible on celluloid. Steven Spielberg's *Jurassic Park* was able to mix live acting with digitally produced dinosaurs so spectacularly that the join was imperceptible. This is good news for the box office but it has its disadvantages. Imaging has come such a long way since the Cottingly fairies that it is now impossible to tell a doctored photograph from an undoctored one.

In which case we all could have been taken in many times without even knowing it.

CRYING WOLPE

In 1992, shortly before the launch of the doomed NASA probe *Mars Observer*, Congressman Howard Wolpe (D-Mich) claimed to have discovered an official two-page document titled "Suggestions for Anticipating Requests under Freedom of Information Act." The document dealt with ways that NASA could circumvent this act and thus withhold from members of the public information which by law they were entitled to see.

Wolpe wrote to Admiral Richard Truly, then head of NASA, saying:

> This NASA document instructs governmental employees to: 1, rewrite or even destroy documents to "minimize adverse impact"; 2, mix up documents and camouflage handwriting so that the documents' significance would be "less meaningful"; and 3, take steps to "enhance the utility" of various FOIA [Freedom of Information Act] exemptions.[2]

Soon after Admiral Truly began his own investigation of this matter he was sacked by President (and former CIA director) George Bush, and replaced by Daniel Goldin who, as we saw in part 1, has a background of secret operations experience. No investigation into NASA's allegedly routine efforts to circumvent the Freedom of Information Act has since been authorized. All this was done, comments McDaniel,

> apparently not to confound enemy spies, but to make it difficult for private citizens, or agencies, or Congress, or the press, to obtain information to which they have a right under the Freedom of Information Act.[3]

With regard to the forthcoming *Mars Observer* mission, McDaniel expressed doubts that NASA would honestly share all new photographic images with the public—particularly any images of Cydonia.[4] Indeed, he pointed out, the agency seemed to have entirely relinquished its control over those images to Dr. Michael Malin, a man known for his implacable hostility to the hypothesis of artificial origins at Cydonia.

MALIN AND OBSERVER

Michael Malin graduated from Cal Tech in 1976 with a doctorate in planetary sciences and geology. From 1975 he had been a member of the technical staff at the Jet Propulsion Laboratory, until he became assistant professor of geology, working his way up to professor in 1987 at Arizona State University. In 1990 he became a research professor and dedicated his time to setting up Malin Space Science Systems (MSSS), of which he is the president and chief scientist.

With the *Mars Observer* mission in 1992–1993, NASA, for the first time in its history, handed responsibility for imaging to a private individual—Michael Malin. Previously, NASA itself had designed, operated, and set targets for its imaging systems. But for *Mars Observer* it contracted MSSS not only to build but also to operate, and be responsible for, all of the imaging done of the Red Planet—including absolute control over any images of Cydonia. As Dr. Malin himself claims:

> No one at NASA has ever attempted to dissuade me from acquiring images in the Cydonia region. No one has ever encouraged me to take such pictures, either, but this is because the choice of areas to photograph has been mine from the start.[5]

We were astonished to learn that even the mission manager at JPL had no authority to tell Malin what to do. But most astonishing of all was the revelation that Malin's *Mars Observer* contract not only gave him absolute authority over where to point the spacecraft and its cameras but also gave his corporation "exclusive control of images obtained from the spacecraft for a period of six months, with no clear statement of accountability."[6]

Understandably this was a state of affairs that worried many AOC researchers. They saw a system ripe for abuse, which appeared almost to have been designed to facilitate the doctoring or suppression of information. For this reason, both before and after the launch of *Mars Observer*, a growing clamor called for Malin's powers to be curbed. For this reason too the AOC lobby continually sought assurances from NASA that the alleged monuments of Cydonia would be reimaged by *Observer* and that the undoctored results would be speedily made public.

To the end NASA never gave such assurances, maintaining a policy that Stan McDaniel describes as "reluctance to assign an appropriate level of priority to rephotographing the AOC objects, coupled with an ambiguous, shifting policy regarding the prompt return of information to the public."[7]

NASA's position was neither a popular nor a defensible one and it seemed to be losing the argument over the mission priorities of *Mars Observer.* The one thing that the public really wanted to know was, would NASA re-image Cydonia and, if so, could we be confident that we would get back original, unaltered pictures?

Or would we get back the reverse of the Cottingly fairy photographs, with evidence of other life removed from the images?

The debate was heating up. As we reported in part 2, it even seemed possible that mission priorities could be changed in response to public pressure. Then, at 6:00 P.M. Pacific Daylight Time on 21 August 1993, all contact with the spacecraft was lost and could not subsequently be restored.

Just like that, just at the crucial moment, *Mars Observer* officially "disappeared."[8]

LOSS

Dr. David Williams at Goddard painted us a picture of the sense of personal disappointment felt by NASA scientists over the loss of *Observer:*

> Well, that was very shortly after I started working here as a matter of fact, and it was pretty devastating—I mean, to have this thing which was right at Mars, and everyone geared up for it, we had spent a lot of time, doing the spacecraft records and experiment records, getting it all set for us to start receiving the data and archiving it, and then it just disappeared. And so it was disappointing to hundreds of people who had invested years and years. I knew some of the people who were investigators on instruments and things for that and it was personally a really bad thing, and even worse for NASA. It was a horrible black-eye; it was a very unfortunate mistake, and it looked bad. It certainly did change, completely turn around, a lot of things about NASA.

Readers will remember the disconcerting fact that this devastating loss occurred during a very risky act—the deliberate switching off of telemetry (contact between *Observer* and Earth). This loss of telemetry was supposedly effected to stop the spacecraft's transmitter tubes from being shocked by the pressurization of the fuel tanks.

> When the valves [that open to allow the helium pressurant to flow to the propellant tanks] actuate, a small mechanical shock wave is set off that travels through the spacecraft's structure and is felt by all the electronic components. . . . One such component is the amplifier tubes in the spacecraft's radio transmitter. The effect is much like causing a hot electric lightbulb to burn out by sharply jostling it while it is on and hot. So, we turned off the radio transmitter to keep it cool so as not to damage it. This is an event that has been done many times previously during the flight of *Mars Observer*. . . . We watched the initial events occur on schedule and the transmitter turn off . . . but we never heard the spacecraft's signal again.[9]

And so, when NASA attempted to regain telemetry, nothing happened. Moreover, the fact that the telemetry had been switched off when the fatal loss occurred meant no record existed of the exact circumstances of the loss (as there would have been with the telemetry on). Many have noted that this communications blackout would have been the ideal window for an act of sabotage—or for a myriad of other scenarios to unfold.

Mars Observer was alone—450 million miles from home. Did it really just suffer an accident, as NASA claimed? Had it found something on Mars that others did not want us to see, necessitating a pulling of the plug? Or was it, and is it even now, orbiting Mars, sending back information . . . to someone?

RESCUE

An official committee, known as the Coffey Board after its chairman Dr. Timothy Coffey (director of research at Washington's Naval Research Laboratory), was set up to investigate the loss of *Observer*. According to Michael Malin, in a note posted on the MSSS website:

> The Coffey Board Report stated that the most probable cause of the loss of communications with the spacecraft . . . was a rupture of the fuel pressurization side of the spacecraft's propulsion system, resulting in a pressurized leak under the spacecraft's thermal blanket. The gas and liquid would most likely have leaked from under the blanket in an unsymmetrical manner, resulting in a net spin rate. This high spin rate

> would cause the spacecraft to enter into the "contingency" mode, which interrupted the stored command sequence and thus did not turn the transmitter on.[10]

Such spinning could also have caused "the main antenna to be torn off. Eventually, because the solar arrays would no longer be pointed at the sun, the spacecraft's batteries would be depleted and unable to power the transmitter."[11]

REBOOT

How hard did NASA fight to reestablish communication? It ought to have fought desperately, yet records show that it delayed a number of vital initiatives for many days—such as mounting a search for *Observer* with the Hubble telescope, for example, and sending the commands to activate the craft's backup computer.

Mars Observer carried two central computers with exactly the same software packages. If the fault had been in the primary computer, then rebooting the secondary computer might have fixed the problem. Even as late as 3 September, however, more than a week after the initial loss of contact with the spacecraft, this obvious remedial action was still being debated.

The reader will recall that *Mariner 9* was shut down for a while in 1971, when it reached Mars in a dust storm. It "hibernated" until the storm was over, and was essentially reprogrammed to start the mapping.

There was no reason why NASA could not have attempted such a move with the second computer onboard *Mars Observer.* Yet inexplicably in the next press release (10 September 1993) the "reboot" option was not mentioned—and never has been since. Did NASA try to reboot the computer? And if not, why not? The secondary computer was placed onboard precisely to fulfill this function. Why not, when you have essentially lost a billion-dollar mission, try this last viable option? NASA's answer at the time was obviously unsatisfactory:

> Analysis by flight team groups indicated greater risk in doing so than is currently deemed necessary in terms of potential effects on other spacecraft telecommunications subsystem components.[12]

So even though the craft was lost, all telemetry defunct, NASA did not wish to reboot the computer because of potential damage to communications equipment! A bizarre state of affairs, considering there was no communication.

One last hope remained of locating the *Observer* and regaining control over it—using a beacon inside a separate component in the craft, the Mars Balloon Relay system. Strangely, no attempts were made to deploy this beacon for a month, when the proximity of Mars to the Sun had resulted in solar interference—essentially camouflaging the 1-watt beacon signal.

SURVEYOR

Within weeks of the loss of *Observer,* NASA announced that it would be sending another orbiter to Mars—a kind of scaled-down *Observer.* This was *Mars Global Surveyor,* which, as we have seen, was launched in 1996 and went into orbit in September 1997. While we were at Cal Tech in summer 1997, we asked Dr. Arden Albee about the *Surveyor* mission and how he reacted to ongoing accusations that NASA did not want to rephotograph Cydonia and the Face.

Dr. Albee was indignant:

> We've always said we were going to do it! I could show you the first description of the *Mars Observer* mission—I wrote it! And it says we're going to photograph the entire surface of Mars.
>
> Now, *Surveyor* will get images of Cydonia all the time, but at low resolution, because the lower resolution camera will cover the planet every day once we get into mapping orbit, so we'll be getting images of Cydonia, but the high-resolution images we will not. We can't predict until we get locked up in our circular orbit.
>
> I will read you a statement that I gave at lunchtime, which I carry for such wonderful occasions.
>
> Question: "Will *Mars Global Surveyor* photograph the Face on Mars?" Answer—my answer, and one to which Malin subscribes, incidentally: "*Mars Global Surveyor*'s camera will provide low-resolution images of the entire surface of Mars. Included in these daily images will be low-resolution images (about 300 meters per pixel) of the Cydonia region, rephotographed on many occasions when the instruments' sur-

face track passes over the region. The camera on this mission does not have the capability to be pointed at specific surface features of interest to scientists. And the mapping orbit from which high-resolution [images will be obtained is designed to allow] viewing of any specific location on the surface of Mars only a few times during the entire mission, within error. Targets within the Cydonia region will be imaged as part of the normal scientific investigation. When the orbital predictions permit, advance notice of these imaging opportunities will be available shortly before they occur, and will be provided over the Internet. After the images are acquired they will also be released over the Internet."

And that's an official project position, an official NASA position, an official Malin position—we'll do our best to take these images, but there's nothing that will satisfy the conspiracy folks.[13]

NASA administrator Dan Goldin is another who has promised to get photos of the Face:

> One of the things we are going to do in our next mission [*Mars Global Surveyor*] is, when the spacecraft goes over the spot, if we have the right pointing, we'll try and take a picture, and scientifically show what we have found.[14]

The reason, Goldin admits, is public pressure:

> I think we have to be somewhat sensitive, especially when we are dealing with government money, to recognize some of the issues that the public has.[15]

UNEXPECTED NEWS

On 26 March 1998 Professor Stanley McDaniel posted on his Web site some much hoped for, but little expected, news:

> This evening I received a welcome telephone call from Glenn Cunningham of the Jet Propulsion Laboratory in Pasadena. . . . Mr. Cunning-

> ham, who heads the *Mars Global Surveyor* project, stated that during April there will be three opportunities to image the area of interest at Cydonia, and that attempts to secure images will be made on each of these occasions.

Fortunately *Mars Global Surveyor*'s positioning and orbit calibration had been completed quicker than expected, and a window had arisen in which the Cydonia anomalies—not officially regarded as scientific targets—could be snapped without altering the main mapping schedule.

In the early hours of 5 April 1998, *Mars Global Surveyor,* 276 miles above the Martian surface, passed silently over the enigmatic and controversial features that had split the scientific community and began rephotographing them. Ten hours later they had been relayed to Earth.

Then, for what seemed an eternity, all waited for the first images to appear.

The silence was broken on 6 April 1998, midmorning Pacific time, as the raw image was posted on the World Wide Web. This long-awaited dark strip of data was an impenetrable mess—and the wait continued for a "cleaner" version of the image via a process of image contrast enhancement that was planned to take "a few hours."

After a number of hours of processing at Malin Space Science Systems HQ in San Diego, the new image was released. To the dismay of many, the words "It's not a face" appeared on Malin's Web site.

"IT'S NOT A FACE."

Amazingly, the *Mars Global Surveyor*'s camera had hit the bull's eye first time and directly pinpointed the Face with breathtaking accuracy.

The new photographic strip was radically different both in acquisition criteria and in content from the original *Viking* frames. As Malin commented:

> The "morning" sun was 25° above the horizon. The picture has a resolution of 14.1 feet (4.3 meters) per pixel, making it ten times higher resolution than the best previous image of the feature, which was taken by the Viking Mission in the mid-1970s. The full image covers an area 2.7 miles (4.4 km) wide and 25.7 miles (41.5 km) long.

The Face was about halfway down the image, and the top right (damaged) corner of the D&M Pyramid was captured.

For a while the supporters of the Face reeled in shock. Was this really the Face? The primary image was unclear and flat, like a series of dunes and ridges encircled by a lozenge of material like a racetrack.

In this image the Face's noble features had been reduced to scars, but it had been a speedy processing, and much of the detail, it soon became apparent, had been bleached out in an attempt to refine the inscrutable primary image. By 5 P.M. that evening further work had been done on the images by Malin Space Science Systems: the image of the Face had been fleshed out and oriented so that it lay at the same angle as the original *Viking* frames.

But still, this was clearly not the Face that the AOC researchers had predicted we would see under high-resolution photography.

McDaniel's reaction was subdued—he said:

> The two "eyesockets" are quite clear, as is the "headdress" or "helmet" feature encircling the object. The small projection on the left cheek appears to be what produced the feature called the "teardrop" in the *Viking* images. There is a face-like appearance, but the overall impression, except for the regularity of the "headdress" feature, is of a natural formation. . . . My initial guess is that the low resolution of the *Viking* images, plus the particular lighting conditions, were what produced the remarkably face-like appearance in the images we are familiar with. On the other hand, there is a sufficiently face-like appearance here as to make the hackles rise. Is it an eerie natural formation, or a heavily eroded intentional sculpture?

He added, in an SPSR press-release:

> In 1976 officials made a snap judgment that the Mars "Face" was "natural" within three hours of receiving the images from Mars. Many of their premature claims turned out to be mistaken. With the arrival of new images from the *Global Surveyor,* there will once again be a temptation to make premature conclusions. No one image of the Face will end the controversy because of the two dozen or so other anomalous formations in the region which form the basis of many of our statistical conclusions.

"I HOPE WE'VE SCOTCHED THIS THING FOR GOOD."

In the next couple of days the world media was awash with NASA's "defacing" of Mars. Quotes appeared from experts, such as Michael Carr of the U.S. Geological Survey, saying, "It's a natural formation, I hope we've scotched this thing for good." But this, like Malin's cry of "It's not a face," may prove a little premature.

For far from ending the argument, it has merely reopened the debate and acted as a catalyst to the controversy.

"IT'S A FACE!"

Richard Hoagland, for one, felt justified in ignoring NASA and Malin's announcements and proclaimed, "It's a face!" There was also a certain logic in other claims that a well-weathered sculpture would actually look less like a face the closer one got. Doubts were certainly beginning to creep in. . . .

Some pointed out that the Face had been photographed in the early morning on 5 April, and yet it waited until 9 A.M. on the 6th to be analyzed, lying apparently untouched in the Project Data Base all night until the start of the next working day, time enough, some might say, for the images to have been altered.

Strangely, it was the first hurried image of the Face that NASA released to the press, the image most unrepresentative of the true form of the landscape, and the image most likely to look incongruous when compared with the *Viking* photos.

The press made little mention of the research of the SPSR, and in many cases failed to mention that the Face was just one feature among many anomalous structures at Cydonia—and as such was not even the strongest case for artificiality. Instead it concentrated on a gleeful debunking of UFO enthusiasts and conspiracy theorists who, it rightly predicted, were unlikely to be dissuaded by the new evidence.

And yet as it stands, the Face is still anomalous—as McDaniel says, it may not be a face—"but what is it?" Many features found by computer enhancements of the original *Viking* frames prove to have been correct, such as the "eyeball" discovered by DiPietro and Molenaar, and the bilateral stripes above the eyes found by Carlotto. Even if these are merely natural, if strange, it proves that other features detected by digital enhancement elsewhere in Cydonia are also likely to exist in actuality, such as the details of the fort, the mound alignments, and the angles of the D&M Pyramid.

However, because it was the Face that had attracted the first attention to Cydonia, its "unmasking" has seemingly destroyed the artificiality hypothesis for many in whose eyes it was, albeit wrongly, the linch-pin on which the whole artificiality argument stood. But we must wait for more detailed pictures of the other enigmatic objects in Cydonia before we can even begin to write off the artificiality hypothesis.

It may well turn out that in trying to lay the ghost of the Face to rest all NASA has succeeded in doing is creating a martyr. Certainly there are signs of a rising tide of dissent against the agency's insistently "natural" interpretation. On 14 April 1998, for example, the following comment from the astronomer Dr. Tom Van Flandern of the U.S. Naval Observatory appeared on Hoagland's Web page: "In my considered opinion, there is no longer room for reasonable doubt of the artificial origin of the Face mesa, and I've never concluded 'no room for reasonable doubt' in my thirty-five-year scientific career."

VALIDATION PERIOD

One issue that has been continually raised in this debate is whether we can be sure, in light of the Wolpe accusation and the Brookings report, that what we are seeing, and will continue to see, in the *Global Surveyor* images, is the whole undoctored truth. Doubts were being aired of the authenticity of the *Global Surveyor* "Face" image within hours of its release, in part due to its difference from the *Viking* images, and in part from the tardiness of its delivery to the public.

This "tardiness" added up to no more than a few hours, explained away by NASA as being due to the reception of the data during the "graveyard shift" when the camera operators were home in bed. Given the fuss made over a handful of lost hours, however, it is no wonder that many were perturbed by the six-month "validation" clause, which, as McDaniel explains, was part of Dr. Malin's contract:

> For some time now, we have been told that the private contractor for the camera onboard, Malin Space Science Systems in San Diego, California,

> has a proprietary period of six months during which it need not release data. On persistent inquiry, I found out just a few weeks ago that now NASA claims there is no such proprietary period—there is instead, they say, a "data validation" period of up to six months. So no matter what it is called, a communication blackout for at least six months after taking any image of Cydonia could take place. Meanwhile, NASA may release images of Cydonia in near real time, but at low resolution from the mapping cameras, essentially useless for the study of the Mars anomalies.[21]

It's easy to see from such pronouncements why many who are interested in the unfolding drama of the "anomalies" tend to regard Dr. Malin as the villain of the piece—a shady background figure, wielding the power to change our entire worldview with a swing of his camera (or at any rate of the spacecraft it is attached to). And yet the man himself has remained invisible, inscrutable, a tabula rasa upon which to project all our Orwellian nightmares—the faceless face of Big Brother NASA.

On 12 December 1997 we contacted Dr. Malin to offer him a chance to give his side of the story. We expected no reply. But the next day, 13 December, we received a four-page e-mail from him containing detailed responses to many of our questions.

THE WIZARD

In *The Wizard of Oz* there is a scene in which Dorothy and her companions reach the Emerald City to find the eponymous Wizard as a threatening, disembodied, thundering voice. Yet Toto the dog pulls back a curtain to show that this is all mechanical trickery performed by a very human wizard indeed.

Communicating with Dr. Michael Malin, the wizard of Malin Space Science Systems, felt a bit like that. Because despite all our expectations he came across as a very human being—intelligent, candid, and humorous.

After reading what he had to say we frankly find it difficult to see him as a villain, and we have begun to suspect that he might really be just a victim of his own consistency. It is as if people's frustration at the scientific world's conservatism and resultant failure to examine the Cydonia question properly have been projected onto the "faceless" Malin for the simple reason that the process of re-imaging Mars, and therefore the Cydonia anomalies, lies in his hands. And the latter was something that, until the surprising rephotographing in April 1998, he had no special plans to do.

Malin forbade us to print his responses to our questions verbatim and seemed concerned that whatever he said would somehow be twisted by us and used against him in an argument that he considers as absurd as it is futile. This is one reason he has kept a low profile—believing that as his responses are usually rejected or claimed to be untrue then it is just a waste of time to reply at all.

CATCH-22

We pressed Malin on the issue of capturing new images of the Face. He answered, as we had expected, that the camera cannot be independently pointed, and that it would be difficult to plan to hit a small target of, say, a few kilometers across.

Time has proven him overcautious here, for as we have seen, when it came to the crunch, Malin was able to target the Face with prodigious, pinpoint accuracy on his first attempt. He added, somewhat prophetically, that even if he did succeed in getting a good image of the Face he thought it very unlikely that the AOC researchers would be satisfied.

As for the epoch-making importance of such a discovery—did he not think it was worth expending the effort, just in case?

The answer was a firm no. Malin said that he considered the probability of the Cydonian anomalies being unnatural as too low to justify the time and money that would be required to investigate them thoroughly.

We remembered David Williams at Goddard telling us that each NASA mission is strictly and tightly funded with a number of set tasks to complete—all of which usually have to be proposed, seconded, and put through numerous selection committees before they get the go-ahead. A five-minute experiment onboard such a probe can be the apogee of a scientist's working life. With this in mind we can easily understand why Malin has no spare time to "follow a whim" such as the Face on Mars. Nor does the fact that the Face has been reimaged suggest any change in his position. Cydonia was only given a chance at re-imaging because of the development of unforeseen spare time between aerobreaking and mapping. Moreover, the re-imaging was undertaken to satisfy public, not scientific, demand. Had this opportunity not arisen, then it is doubtful that the Face would have been specifically targeted at high resolution.

But it is precisely this lengthy selection process that the AOC researchers find so invidious. There are no scientists within NASA approaching committees to fund their kind of research—and since the tragic loss of the *Challenger* shuttle and *Mars Observer,* money is tighter than ever. It seems that NASA can only afford to send a mission to investigate the entirety of the Cydonian anomalies fully and systematically if there is undoubted proof of artificiality. This is a catch-22, say the AOC researchers, because unambiguous proof, one way or another, is only likely to be obtained by precisely such a mission. And, given the latest damning criticisms of the Face based on the *Mars Global Surveyor* image, such an investigation seems even more unlikely than before.

DELICATE ISSUES

In our questions to Dr. Malin we turned next to the delicate issue of the loss of the *Mars Observer.* What did he make of widespread allegations that he himself had pulled the plug—or even that images were being secretly relayed back as we spoke?

Malin's reply was bitter and direct. The loss of *Observer* had been a horrible disaster for him, forcing him to fire half his staff and to move those remaining into temporary buildings. If he had sabotaged his own mission, he argued, where were the benefits? While the AOC researchers lined their pockets from writing and lecturing about such issues, he had suffered the loss both personally and financially. He then turned the question back on us: How would we respond to such cruel allegations?

As for the six-month validation period, Dr. Malin argued that this was not in any way sinister but simply a practical necessity when operating on such a small budget, allowing time to process all the images into a workable format. There were just no resources to assemble a massive team to do this instantly, as the information came in. Press releases would show important results quickly, but that was a different process—one not budgeted for in Malin's contract. The rest of the hard slog of image retrieval would take most of the six months, and whatever time was left over would be used for evaluation and interpretation.

COVER-UP, OR JUST MONEY?

The whole issue, in other words, seems to boil down not so much to secrecy as to money.

And this, in the final analysis, is why Malin says that he is so unhappy about the Face controversy—and also, more generally, about the search for biological life on Mars. In the *Viking* missions, he reminded us, looking for life on Mars had led absolutely nowhere at great expense. Money that could have been spent on bona fide scientific investigations—for example, assessing the possibilities for future human habitation of the Red Planet—had been, in his opinion, squandered on biological experiments that were insubstantial. He sees the quest for life as little more than an ego trip for scientists wishing to be the first to make a sensational discovery.

Malin, it seems, is content just to be a scientist, not a celebrity—a point that rings true in the light of his reluctance to talk on this issue, and his failure to exploit his situation for personal financial gain. As he told us, he could earn a fortune were he to be the man who found life on Mars.

Portraying himself as a conscientious scientist who knows the limitations of NASA's budget, he says that he simply wishes to be pragmatic and get the best out of what he has rather than tilting at windmills. This is a cautious approach, and one that could be faulted for its lack of pioneer spirit—but NASA is not endowed with limitless funds. Realistically this means that Malin, who knows from personal experience that the space program is financially flimsy, is effectively constrained from the start.

On balance it is our conclusion that NASA is not really a secretive cabal like the CIA and the FBI but a body made up of scientists and enthusiasts whose zeal for their subject is as admirable as it is infectious. A pervasive sense of something being "covered up" does, nevertheless, infect the organization, but if there is a conspiracy involving the Mars monuments and other "extraterrestrial" issues we are fairly sure that it is not at grassroots level—where there would be great excitement and interest if evidence of extraterrestrial life were ever to be found.

In any rational appraisal of the whole problem it should not be forgotten that NASA's own enthusiasts are kept in check by government and must operate within parameters established by government. Moreover, as we have shown, the agency has been closely linked throughout its history to national defense and security. Indeed, it must be remembered that documents like the Brookings report advise that as far as possible *even the scientists themselves should be kept in the dark* if evidence of extraterrestrial life is ever confirmed.

So we cannot entirely rule out a high-level conspiracy—one way over the heads of ordinary scientists but thriving on their dogmatic, narrow-minded, and unadventurous attitudes and sustained by ferocious competition for scarce resources. Even in a conspiracy such as this, however, it might be difficult to prevent leaks of information about Mars emanating from our ancestors in the distant past.

Far-fetched though it may sound, we will show in the next two chapters that there are merits to this scenario.

16

Cities of the Gods

REMEMBER latitude 19.5 degrees north—the landing site in July 1997 of the tetrahedral *Mars Pathfinder*—and the discovery of the mathematical values *phi, pi, e,* and *t,* as well as *sqrt* 2, *sqrt* 3, and *sqrt* 5, in the pyramids and mounds of Cydonia? Several AOC researchers do not believe it can be an accident that identical geometry (and identical latitude preferences to within 2 arc minutes—that is, two-sixtieths of a degree) are found at several archaeological sites on Earth.

In the Valley of Mexico, ancient Teotihuacan, "the place where men became Gods," sprawls near latitude 19.5 degrees north, close to modern Mexico City. A wonder of antiquity—of unknown origins and of uncertain age—its four-kilometer-long Way of the Dead is overlooked by three monstrous pyramids: the Pyramid of the Sun, the Pyramid of the Moon, and the Pyramid of Quetzalcoatl.

In 1974 Hugh Harleston, Jr., a civil engineer obsessed with Meso-America since the 1940s, presented a controversial and revolutionary study of the city of Teotihuacan at the forty-first International Congress of Americanists.[1]

After thirty years of calculation, and more than 9,000 on-site measurements, he stumbled across the hitherto unknown system of measurement used at Teotihuacan—which he named the STU, the Standard Teotihuacan Unit.[2] This unit is equivalent to 1.059 meters. John Michell, an authority on ancient metrology, has this to say about the STU:

> [Harleston] also recognized the geodetic significance of that unit; 1.0594063 meters is equivalent to the "Jewish rod" of 3.4757485 feet, the same unit which represents the width of the Stonehenge lintels, a six-millionth part of the Earth's polar radius, and one part in 37,800,000 of its mean circumference.[3]

THE CODE

Harleston found that the measurements of structures in Teotihuacan, and also the distances between specific structures, are governed by a distinct sequence of numbers in STUs—notably 9, 18, 24, 36, 54, 72, 108, 144, 162, 216, 378, 540, and 720 STUs. Thus, for example, the length of one side of the Pyramid of the Sun at the base is 216 STU, the length of one side of the Pyramid of the Moon at the base is 144 STU, and the center of the Pyramid of the Sun lies 720 STU south of the center of the Pyramid of the Moon.

What is interesting about this sequence of numbers, as science historians Giorgio de Santillana and Hertha von Dechend have shown in their masterwork, *Hamlet's Mill,* is that it recurs continuously in ancient myths and sacred architecture all around the world.[4] These authorities have also demonstrated that the sequence is derived mathematically from an astronomical phenomenon known as the precession of the equinoxes.

To summarize briefly, it is sufficient to remind the reader that there is a minute wobble on the axis of the Earth and that this wobble has a cycle of 25,920 years. Since the Earth is the viewing platform from which we observe the stars, it is inevitable that these minute changes in Earth's orientation in space will alter the apparent orientations of stars as they appear when viewed from Earth.

The best-known effect of precession is observable on the spring equinox, 21 March in the Northern Hemisphere, and manifests as an extremely slow revolution of the twelve zodiacal constellations against the background of which the sun is seen to rise on that special day. This revolution proceeds at the rate of one degree every 72 years (and thus 30 degrees in 2,160 years). Since each of the twelve zodiacal constellations has traditionally been allocated a 30-degree segment of the ecliptic (the perceived annual path of the Sun), it follows that each will "house" the sun on the equinox for a period of 2,160 years (12 × 2,160 = 25,920 years, the complete precessional cycle).

These numbers and calculations form the basic ingredients of an ancient code. Let us call it the "precessional code." In common with other esoteric numerological systems, the code is one in which it is permissable to shift decimal points left or right at will and to make use of almost any conceivable combinations, permutations, multiplications, divisions, and fractions of certain *essential* numbers (all of which relate, precisely, to the rate of the precession of the equinoxes).

The ruling number in the code is 72. To this was frequently added 36, making 108, and it was permissable to divide 108 by two to get 54—which could then be multiplied by ten and expressed as 540 (or as 54,000, or as 540,000, or as 5,400,000, etc., etc.). Also highly significant is 2,160 (the number of years required for the equinoctial point to transit one complete zodiacal constellation). This could be divided by ten to give 216, or multiplied by ten and factors of 10 to give 216,000 or 2,160,000, etc., etc. The number 2,160 was also sometimes multiplied by two to give 4,320—or 43,200, or 432,000, or 4,320,000, and so on.

We have demonstrated in other works that the code occurs in the architecture of Angkor in Cambodia and the pyramids of Giza in Egypt.[5] At Giza we have shown that it is the key that unlocks a precise mathematical scale model of the northern hemisphere of the earth. Thus, if you multiply the height of the Great Pyramid by 43,200 you get a precise printout of the earth's polar radius, and if you multiply the measurement of the base perimeter of the Pyramid by the same figure you get a precise printout of the earth's equatorial circumference.[6]

The same sort of thing happens at Teotihuacan. For example, as Harleston's survey demonstrates, the distance in STUs along the boundary buildings of the Pyramid of the Moon—378—and the distance in STUs of one side of the base of the Pyramid of Quetzalcoatl—60—produce interesting numbers when multiplied by 100,000. The former gives the circumference of Earth and the latter the planet's polar radius.[7]

Harleston established his data by 1974, two years before the first *Viking* photographs of Cydonia were taken. We were therefore interested to learn another mathematical secret revealed by his measuring survey: the builders of Teotihuacan went out of their way to relate structures to one another through ratios of *pi* and *phi* and *e*.[8] Harleston's conclusion was that they must therefore have possessed knowledge comparable to that of modern-day geographers and astronomers:

> Here was a design whose dimensional configurations provided accurate universal mathematical and other constants with a minimum of shared points . . . laid out . . . to incorporate the values of *pi, phi,* and *e.* Perhaps the pyramid complex was an intended hint to latecomers to expand their consciousness for a clearer view of the cosmos and of man's relation to the whole.[9]

IT KNOWS WHERE IT IS . . .

The reader will recall that the D&M Pyramid at Cydonia was shown by Erol Torun to be located at latitude 40.868 degrees north, the tangent of which is the equivalent to *e/pi.* Thus he concluded that it was intelligently founded on that latitude, and self-referencing. Harleston was to discover something very similar when he measured the Pyramids of the Moon and Sun at Teotihuacan. In brief, the angle of the fourth level of the Pyramid of the Sun is set at 19.69 degrees—the exact latitude of the pyramid itself (which stands at 19.69 degrees north of the equator).[10] It is, therefore, a self-referencing monument that makes use of geometry to tell us that it "knows where it is"—that is, it knows its own latitude—just as the D&M Pyramid does. What's more, the angle of the corresponding level of the Pyramid of the Moon, the fourth level, is set at exactly the *t* constant of 19.5 degrees so favored in the overall design of Cydonia.[11]

Such figures have suggested to some researchers that Teotihuacan may contain a "message"—perhaps identical to that of Cydonia—based on tetrahedral geometry and the *pi, phi, e,* and *t* constants. Nor is Teotihuacan the only object of such exotic suspicions.

MEGALITHOMANIA

Stonehenge, the great ring of megaliths that dominates Salisbury plain in Wiltshire, England, is thought to have been built between 2600 and 2000 B.C.—although with some much earlier, and some slightly later stages. It is not our purpose to embark on an exploration of this most intriguing of sites, whose astronomical and geodetic qualities would call for a book in themselves, but to review some of the comparisons to Cydonia that the Mars researchers have made.

According to Carl Munck, for example:

> The very angle off true north of [Stonehenge's] famed northeast avenue (as opposed to the current azimuth of the rising solstice sun) is, astonishingly, another key "Cydonian angle"—49.6 degrees. Identical not only to a key theoretical "tetrahedral" angular relationship [to within 0.2 arc seconds] . . . but also identical to another specific angle, expressed twice in the internal geometry of the D&M Pyramid itself![12]

This angle is none other than *e/pi* when expressed in radians.

Avebury, also in Wiltshire—dated to approximately the same period as Stonehenge, perhaps even a little earlier—is the largest stone circle in the world, containing a village and two smaller stone circles within its environs. What levels of coincidence are called for to explain the fact that centers of the two inner circles of Avebury are offset from true north at an angle of 19.5 degrees?[13]

Because the angle of 19.5 degrees has no intrinsic meaning save as *t*, the circumscribed tetrahedral constant, we can only assume that its repeated reappearances in ancient and sacred terrestrial sites must be deliberate and must be derived from sophisticated tetrahedral geometry. But how do we explain the fact that it also occurs repeatedly in the "monuments" of Cydonia, millions of miles from Earth, on the ruined Red Planet, Mars?

NUMBERS ON THE NILE

We have seen that what seems like a specific mathematical code involving tetrahedral geometry and numbers derived from the precession of the equinoxes lies hidden in the measurements of many of the world's ancient sites. Paramount among these sites is Egypt's remarkable Giza necropolis, containing the Great Sphinx and the Pyramids of Khufu, Khafre, and Menkaure.

Mars, the Red Planet: At present a bleak and barren hell, but mystery surrounds its past. *(NASA)*

The northern polar cap of Mars is formed from frozen water and carbon dioxide—perhaps the remnants, scientists believe, of a once carbon-dioxide-rich Martian atmosphere that would have warmed the planet, allowing the now frozen water to flow freely—providing a suitable climate for the formation of life. *(NASA)*

The southern polar cap of Mars consists entirely of frozen carbon dioxide. *(NASA)*

(BELOW) Olympus Mons: At 700 kilometers in width, it is the largest volcano in the solar system—some three times higher than Mount Everest. *(NASA)*

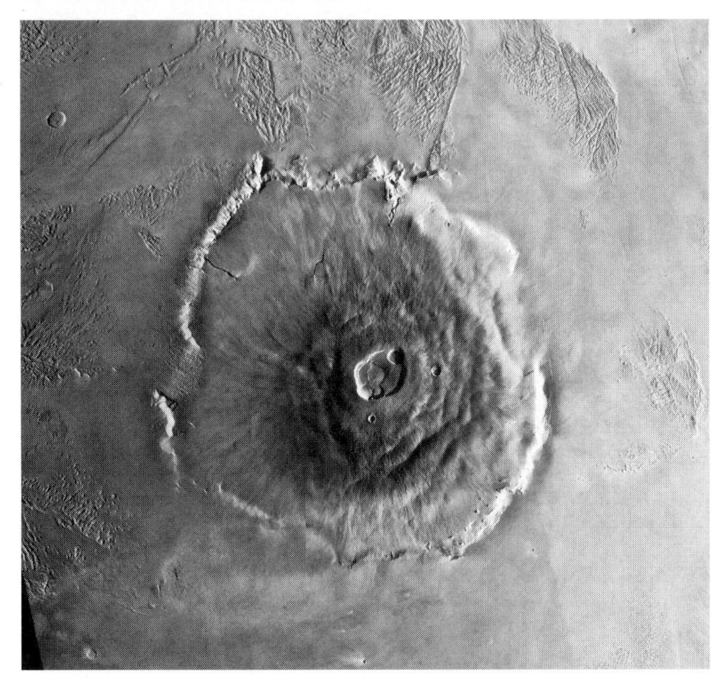

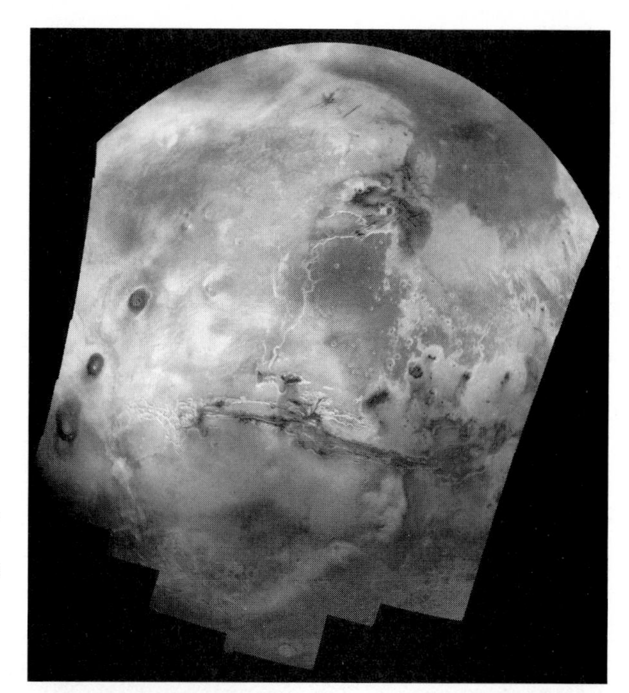

(RIGHT AND BELOW) The Tharsis Bulge surmounted by three gigantic shield volcanoes—Arsia Mons, Pavonis Mons, and Ascraeus Mons—known collectively as the Tharsis Montes. *(NASA)*

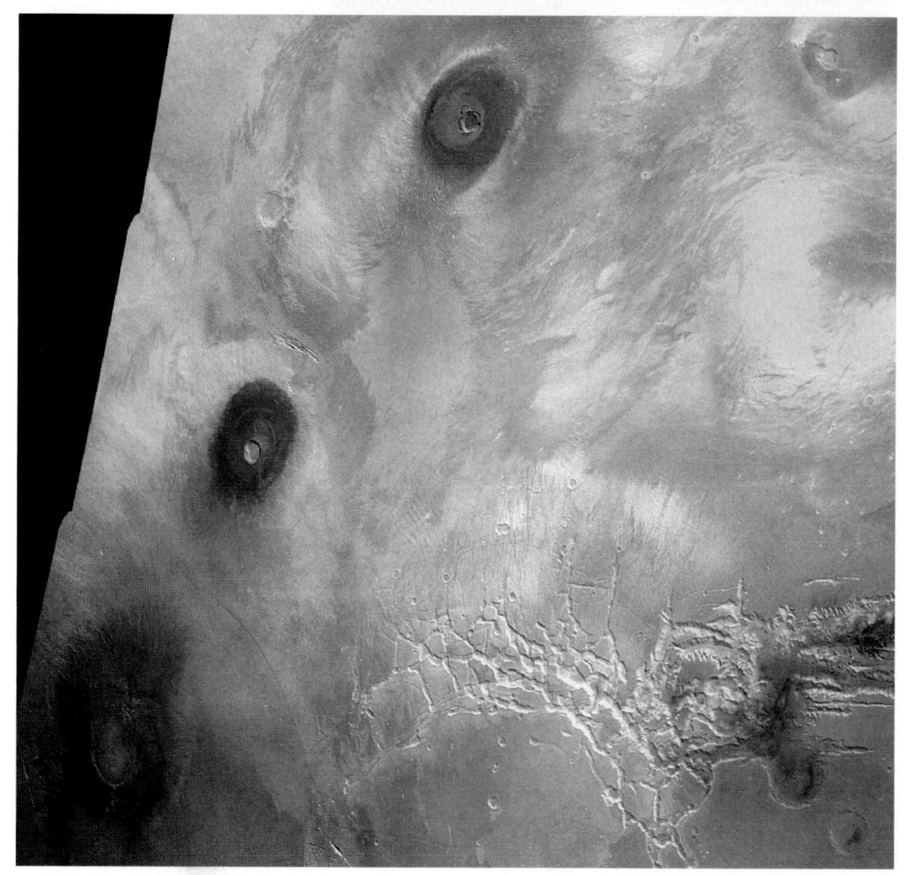

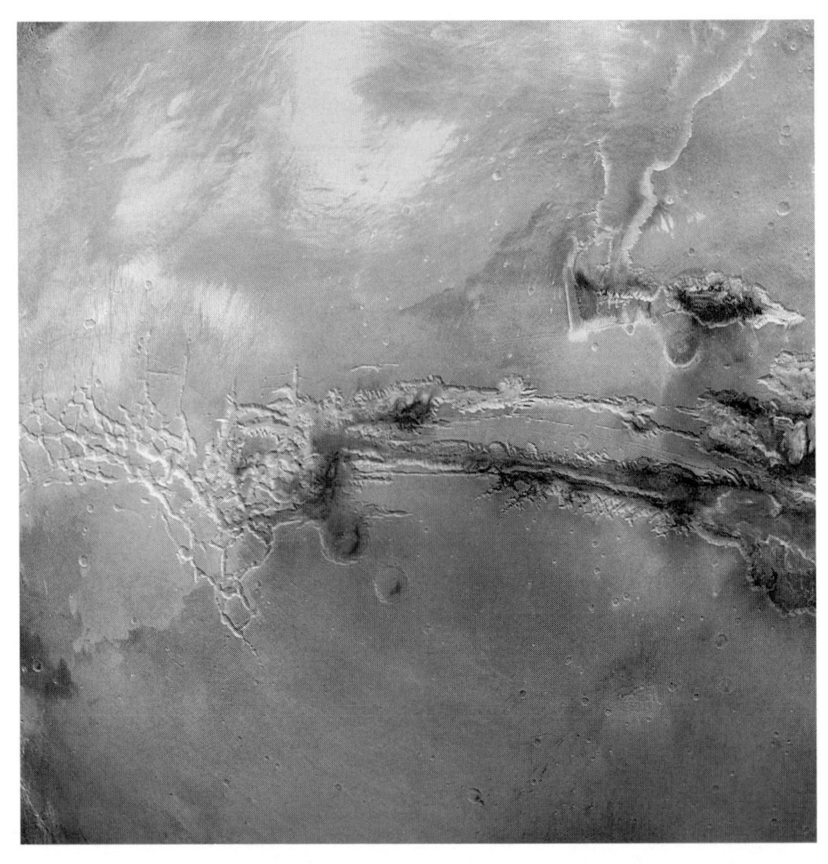

The immense canyon of the Valles Marineris is up to 7 kilometers deep, with a maximum width of 200 kilometers. *(NASA)*

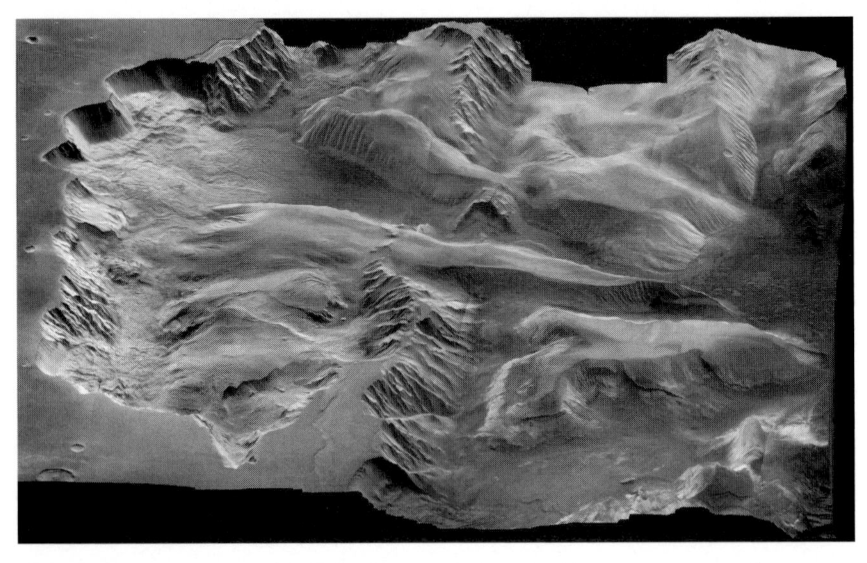

The breathtaking immensity of the Valles Marineris is shown in this computer-generated reconstruction of the Candor Chasm, one of the deepest parts of the Valles. *(NASA)*

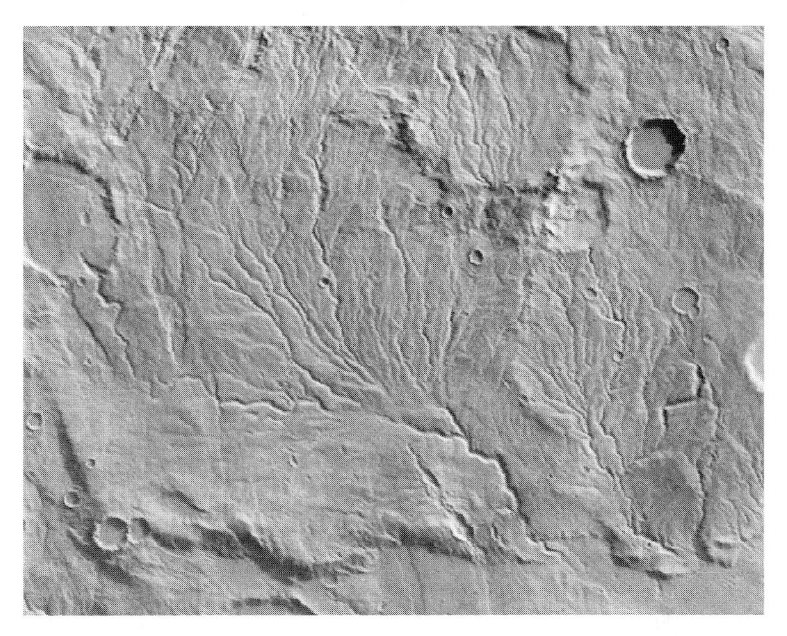

"Dendritic" channels such as these, resembling terrestrial river tributaries, give tangible evidence that barren Mars may once have been as abundant in water as Earth is today. *(NASA)*

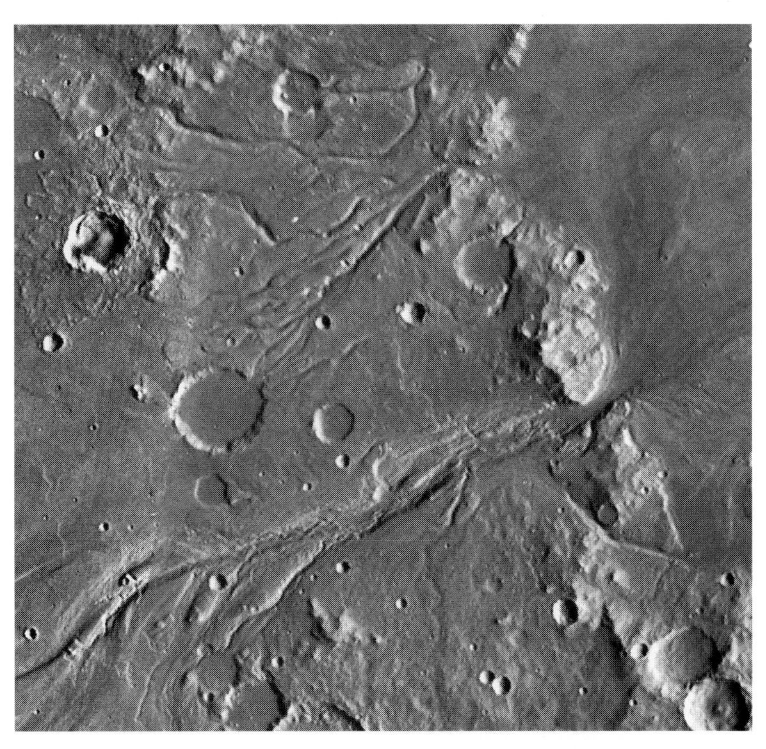

Were these channels in the Chryse Planitia formed from the movement of great bodies of water? *(NASA)*

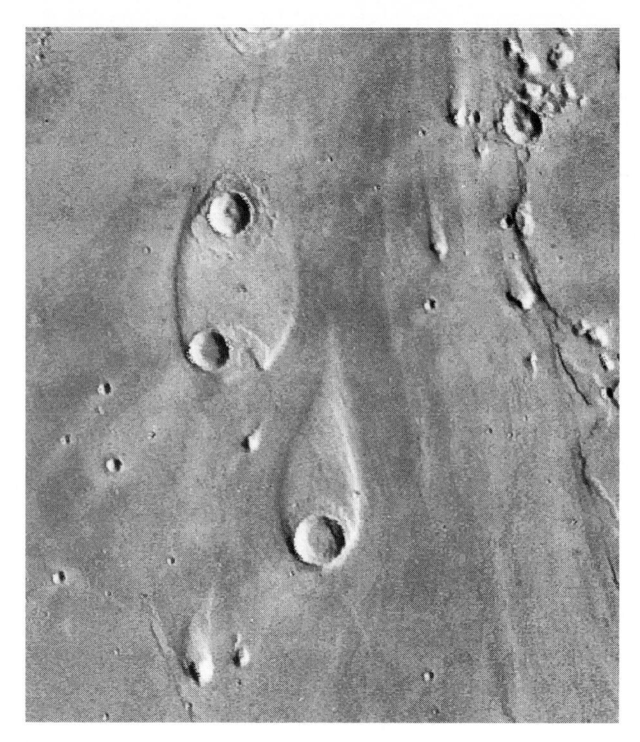

Teardrop-shaped islands in the Chryse Planitia strongly suggest that Mars may once have experienced floods of biblical proportions. *(NASA)*

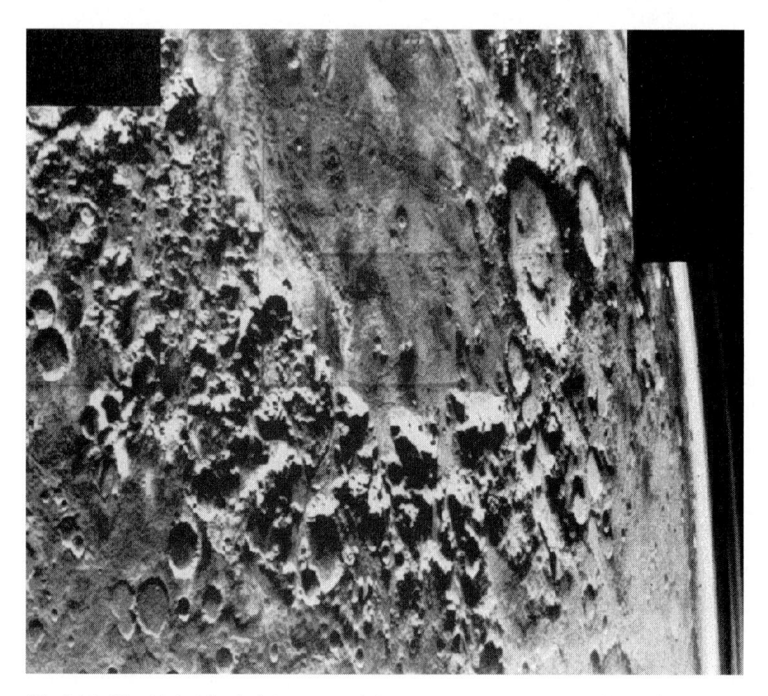

The Isidis Planitia's ridged plain—1,000 kilometers across—was caused by a devastating head-on collision with an object 50 kilometers wide. *(NASA)*

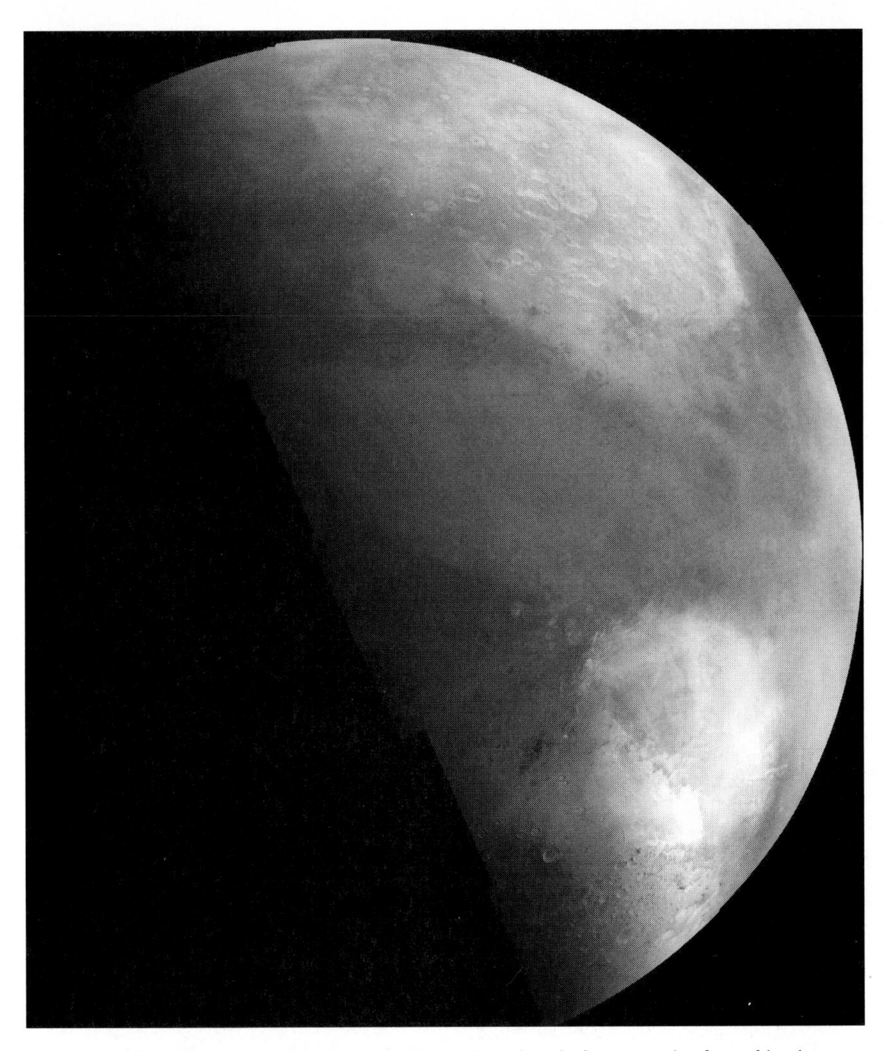

The Hellas Planitia's ridged plain, here swathed in carbon dioxide frost, was also formed by the impact of an object one hundred kilometers in size. *(NASA)*

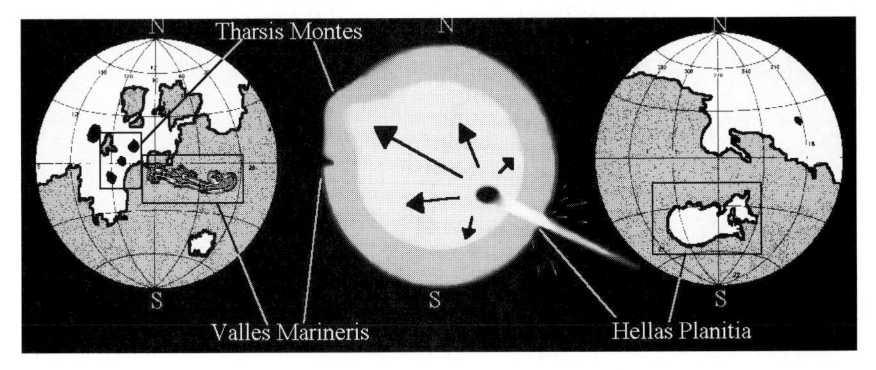

The internal distress caused by the impact of subsequent fragments may have resulted in the formation of the Tharsis Bulge in the opposite hemisphere, causing Mars to burst its seams along one-quarter of its circumference to form the chasm of the Valles Marineris.

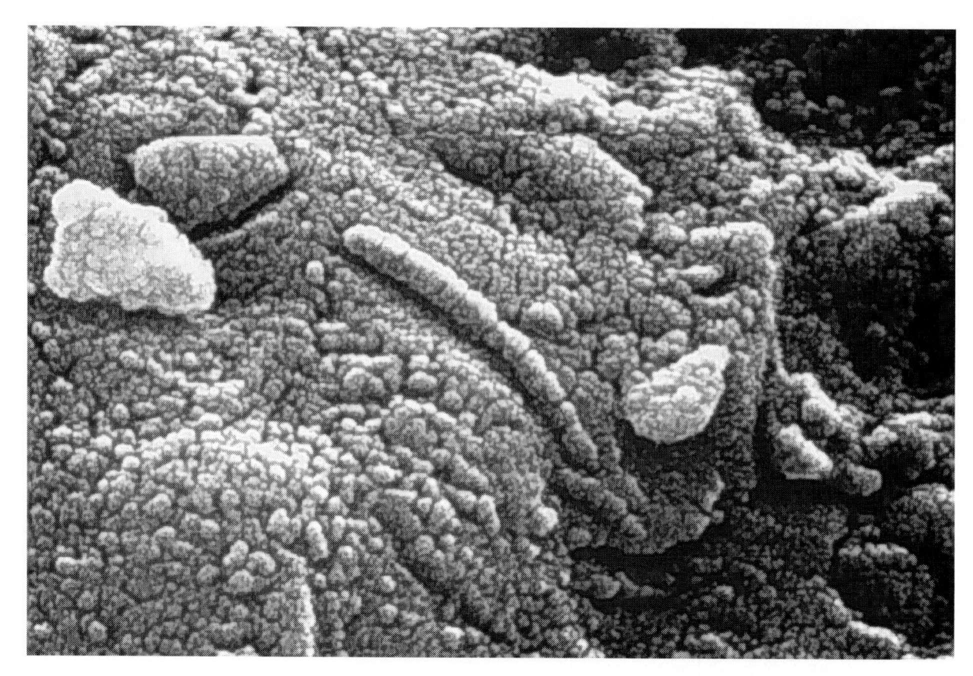

This microscopic image of a Mars meteorite shows possible fossils of bacteria-like organisms found in Martian meteorite ALH84001. *(NASA)*

Phobos—the largest of Mars's two moons. *(Internet)*

The world's first view of "the Face" on Mars as released in 1976 by the Jet Propulsion Lab in Pasadena, California. NASA immediately dismissed it as a "trick of light and shadow." *(NASA)*

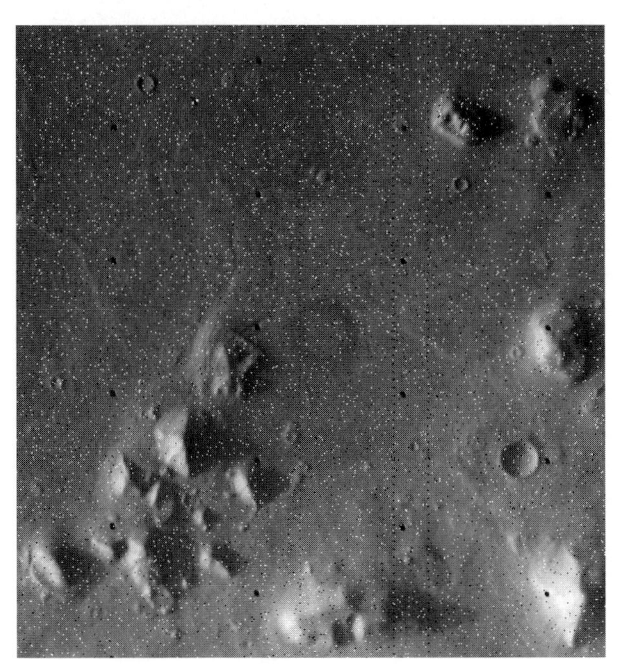

Even before computer enhancement, the haunting image of the Face clearly stares up from Viking frame 35A72. *(NASA)*

Frame 70A13, discovered by Vincent DiPietro and Gregory Molenaar, confirms evidence of frame 35A72 by revealing that the Face is not a trick of light, but actually face-like in form, as this latter frame was taken from a different position and with a different sun-angle from the first. Had it been just "a trick of light and shadow" as NASA claimed, then the Face would have disappeared when viewed from a different position. *(NASA)*

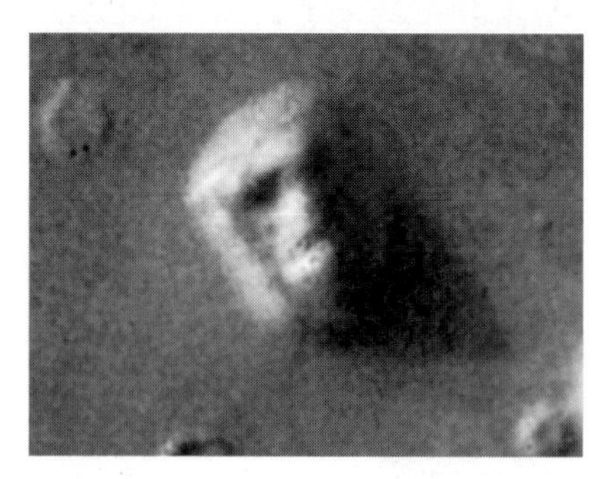

Dr. Mark Carlotto's digitally enhanced image of the Face from frame 35A72. *(Mark Carlotto)*

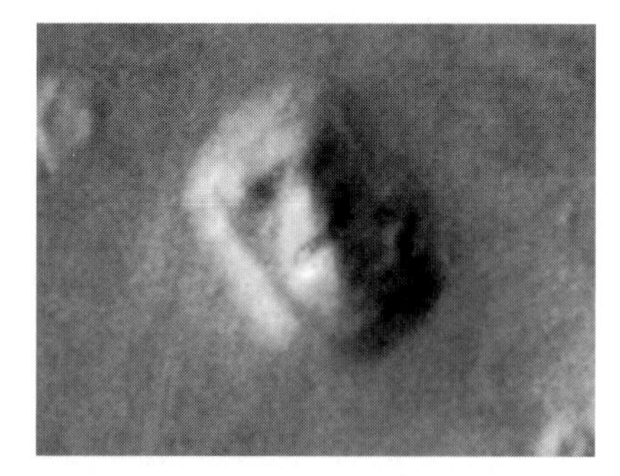

Similarly enhanced image of the Face from frame 70A13. *(Mark Carlotto)*

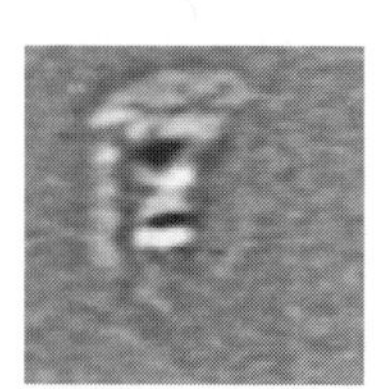

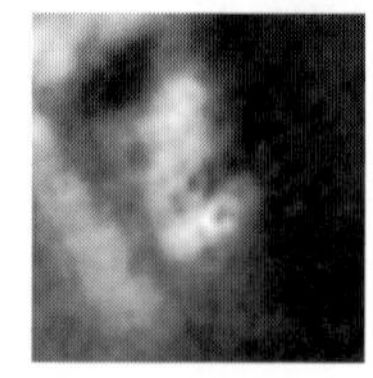

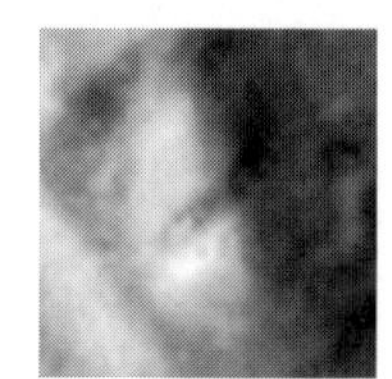

Computer image processing reveals subtle details not visible in the raw Viking images. These include bilaterally crossed lines above the eyes and stripes reminiscent of the Pharaonic *"Nemes"* on the headdress; a tear below the eye; and teeth in the mouth—all of which occur in both frames 35A72 and 70A13, thus reducing the possibility of their being artifacts of image processing. *(Mark Carlotto)*

Naturally occurring "faces" on Earth (note that they only tend to occur in profile, and only work from specific angles); see how the face in the middle image disappears in the right image when photographed from a different angle, whereas the Face on Mars retains its facial characteristics from whichever position it is viewed. *(Mark Carlotto)*

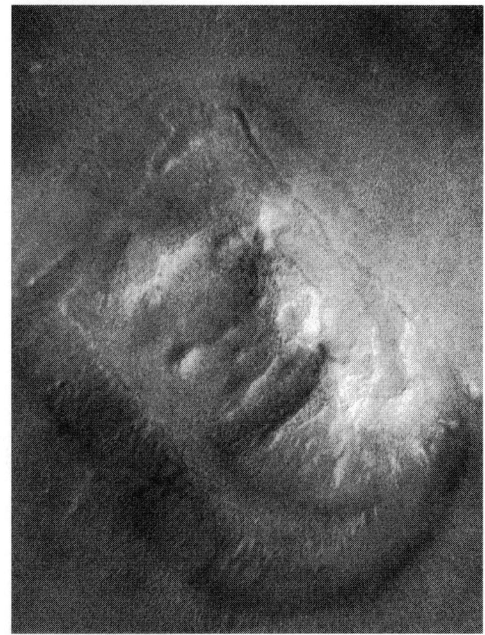

On April 5, 1998 the *Mars Global Surveyor* captured this image of the Face. NASA has tried to dismiss the image as proof that the Face is natural, but the new image has only heated controversy around the Face's artificiality. *(NASA)*

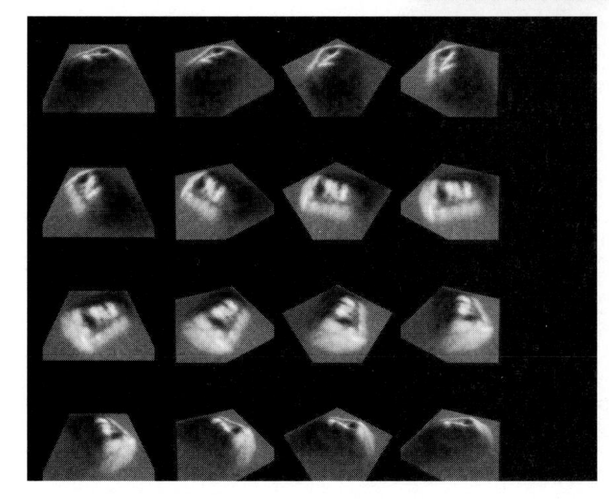

Three-dimensional reconstructions of a 1976 Viking frame, using "shape from shading" technique, of different views of the Face, revealing it to be face-like from all angles.
(Mark Carlotto)

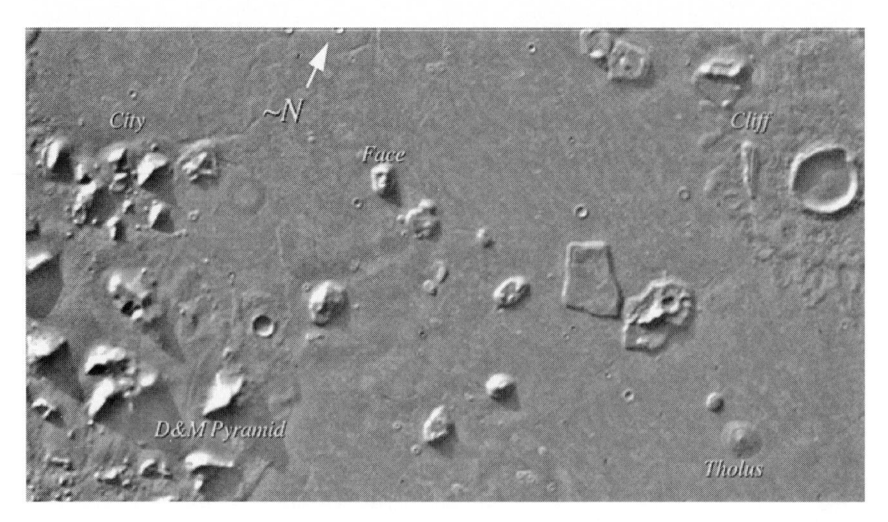

(ABOVE) Dr. Mark Carlotto's digital enhancements of the Viking frames yields a dramatic overview of the weird grouping of the anomalies on the Cydonian plain—including the Face, the City, and the enigmatic D&M Pyramid—much more conducive to an artifical landscape than any freak arrangement of random geological processes. *(Mark Carlotto)*

(RIGHT) A three-dimensional computer-generated overview of the Cydonian anomalies, looking toward the Face from the City. *(Mark Carlotto)*

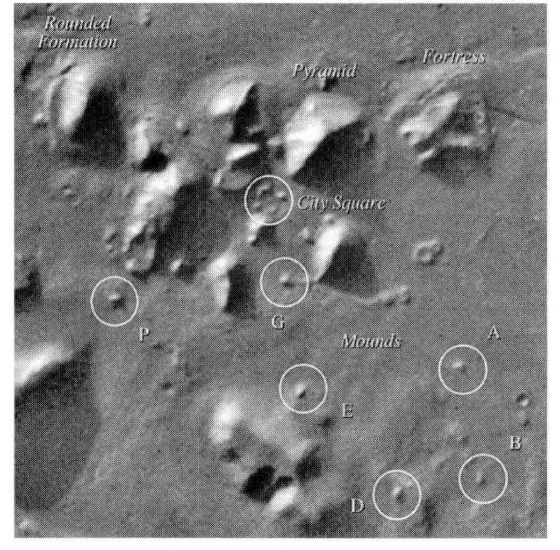

(LEFT) The enigmatic structures known as the City. *(Mark Carlotto)*

(ABOVE) The cliff and crater. Note how the linear structure on the cliff seems unaffected by the ejecta-splash of the nearby crater, as if it was positioned after the formation of the crater, pointing to the fact that this feature is possibly non-natural, as it is not a feature of the original landscape, but a relative latecomer. *(Mark Carlotto)*

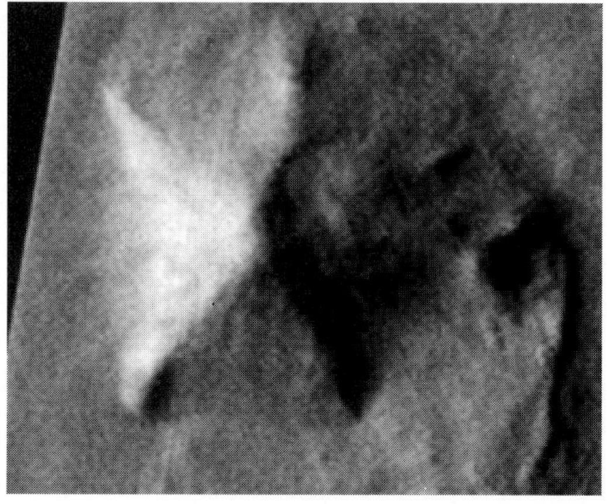

(CENTER) The mysterious five-sided structure known as the D&M Pyramid (named after its discoverers, Vincent DiPietro and Gregory Molenaar) rises some 1,250 meters from the surrounding plain, and is about a thousand times greater in volume than the Great Pyramid at Giza. *(Mark Carlotto)*

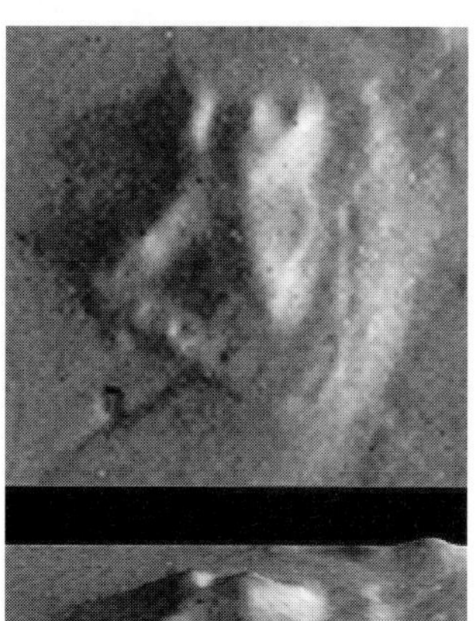

(LEFT) The so-called Fort with its inexplicable angular walls (top). There is no known natural process that can carve straight sides like this on the exterior and interior of an object concurrently with such angular precision. Are we witnessing a new geological phenomenon or the evidence of intelligent design? A computer-generated 3-D perspective viewed from above is seen at bottom. *(Mark Carlotto)*

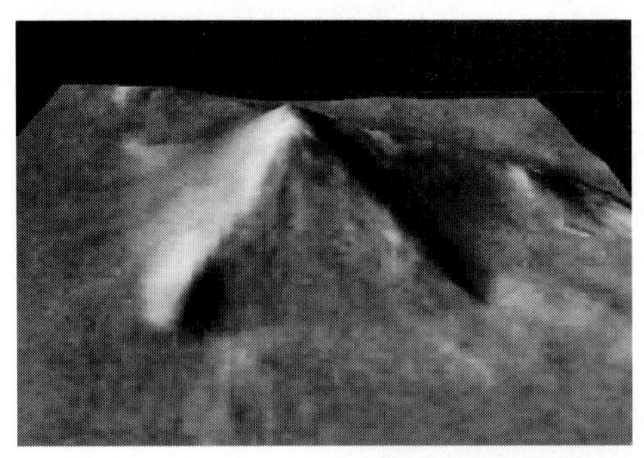

Computer-generated perspective view of the D&M Pyramid showing what some believe to be a tunnel-like entrance to the right. *(Mark Carlotto)*

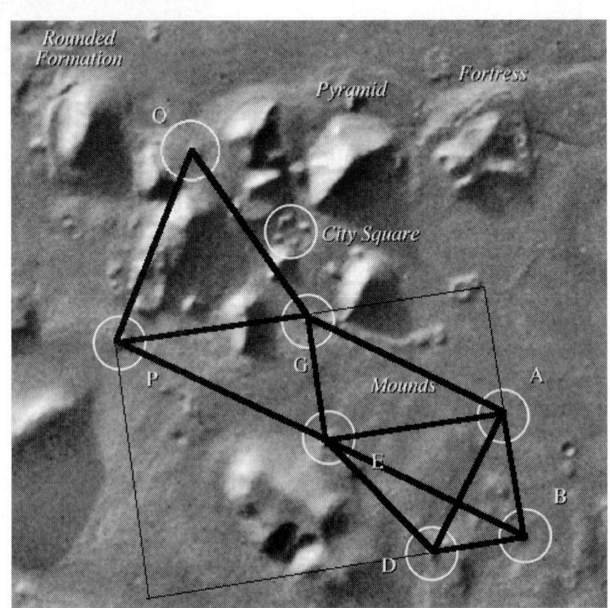

Professor Horace Crater's analysis of the layout of mounds within the City area reveal that their alignment is unlikely to have occurred naturally, as they consist of repetitious patterns of the same basic triangular units expressing meaningful geometrical measurements, whose chances of occuring naturally are astromonically low.

Illustration of *Mars Pathfinder* and the *Sojourner* rover, whose 1997 landing was headline news worldwide, bringing Mars back into the public debate after twenty years of silence. *(NASA)*

On July 16, 1994, the first of 21 fragments of comet Shoemaker-Levi 9 collided with Jupiter. Gene Shoemaker, co-discoverer of the comet, when asked what could be learned from this collision, simply replied, "Comets really do hit planets." *(NASA)*

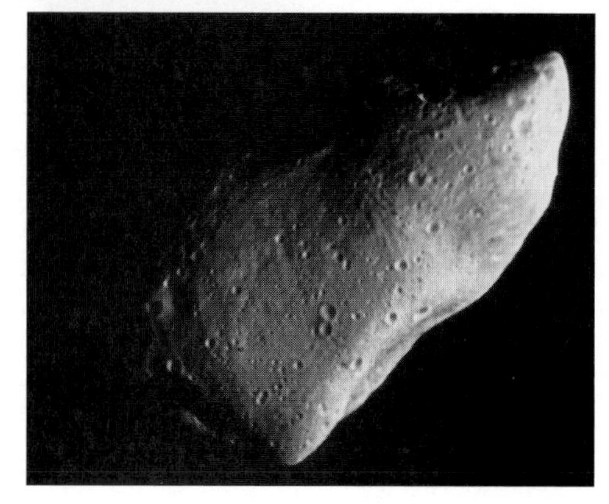

Our solar system is teaming with asteroids, many of which regularly cross Earth's orbit. Here we see Gaspra, a sizeable inhabitant of the asteroid belt that lies between Mars and Jupiter. *(NASA)*

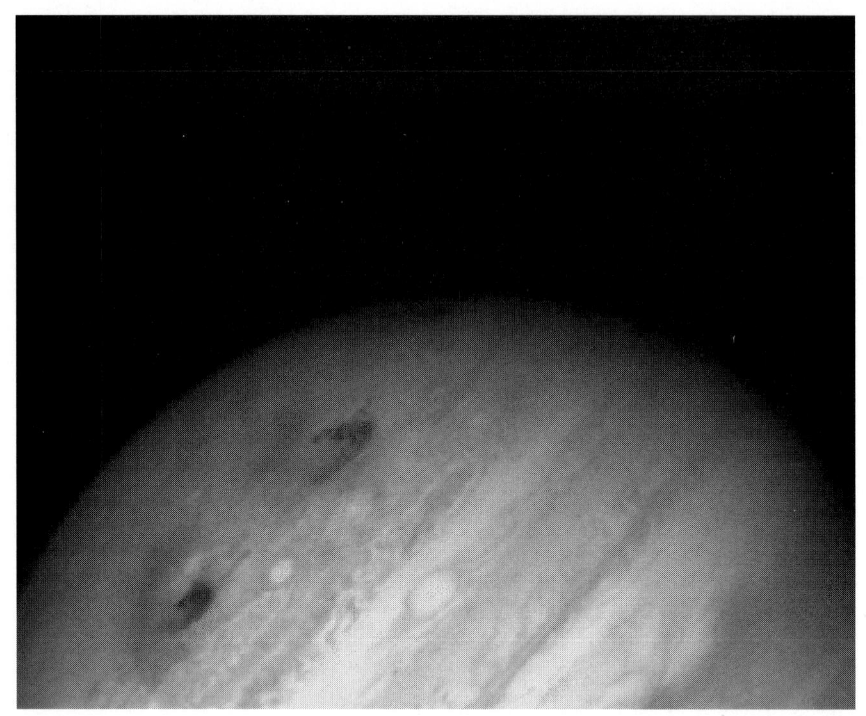

The scars left on Jupiter from the impacts of Shoemaker-Levi 9. The impact ring created by cometary fragment G was larger than Earth. *(NASA)*

Doomsday! Cometary impacts on Earth were most probably responsible for mass extinctions in the past, such as the extinction of the dinosaurs. If Earth was to suffer a large cometary impact in the future, it is entirely possible that all human life would be wiped out in an instant. Are we doomed to share the fate of our red neighbor and become a barren and lifeless hell-world? *(NASA/Don Davis)*

Erol Torun has shown that if we use the apexes of the three pyramids to form a Fibonacci curve (the curve produced within *phi,* the golden section) then the exact location of the Sphinx is dictated by the rectangles that house this curve—indicating that the pyramid builders must have had a good knowledge of *phi.*[14]

Other notable "number games" are as follows:

- The slope angle of the Great Pyramid is 51 degrees 51 minutes 40 seconds. The cosine of this angle is 0.6179, which can be rounded within three decimal places accuracy to 0.618. As the reader will recall the golden *phi* ratio is 1:1.618. The figure of 0.618 is the amount that must be added to 1 to produce *phi.*
- Correct to two decimal places, *phi* is also hinted at by the ratio between the slope of the Great Pyramid and the angle of culmination of the sun at the latitude of Giza at the summer solstice in the epoch of 2500 B.C., estimated at 84.01 degrees (51 degrees 51 minutes 40 seconds; that is, 51.84 degrees, divided by 84.01 degrees, equals 0.617).[15]
- Within the depths of the Great Pyramid, in the enigmatic King's Chamber, is it a coincidence that the wall height plus half the width of the floor produces the measurement of 16.18 royal cubits, again incorporating the essential digits of *phi*?
- Let us return to the Great Pyramid's slope angle and the way in which its cosine generates a figure related to *phi.* We have also seen that there is a relationship between the slope angles of Teotihuacan and the latitude of the site, and between Cydonia's latitude and *e/pi.* Now the Great Pyramid's latitude is 29 degrees 58 minutes 51 seconds. If we round this out to 30 degrees, we will find that the cosine to within one decimal place is 0.865—that is, the tetrahedral ratio *e/pi.*
- The *e/pi* value also seems to be incorporated in the ratio of the Great Pyramid's slope angle (51.84 degrees) to the slope angle of the southern shaft of the King's Chamber (45 degrees). This ratio is again within one decimal place of *e/pi.*
- *Pi* is found in the base perimeter-to-height ratio of the Great Pyramid (1,760 to 280 cubits = 2 *pi*).

A SINGLE UNIFYING THEME

In 1988 in an article in the obscure scholarly journal *Discussions in Egyptology,* the British mathematician John Legon published intriguing data on the siting of the Giza monuments, showing that "the size and relative positions of the three pyramids were determined by a single unifying theme."[16]

These monuments, he pointed out, are

> accurately aligned with respect to the four cardinal points, and the bases are displaced from one another in a formation that meets the requirements of a coherent dimensional relationship. Difficulties with the site chosen for each pyramid also suggest that there must have been some constraint in addition to the usual factors such as ease of construction or architectural setting.[17]

When he drew a rectangle that would exactly enclose the three pyramids, Legon discovered that its dimensions were 1,417.5 cubits east to west, and 1,732 cubits north to south.[18] Within a fractional margin of error these figures are equivalent to 1,000 × *sqrt* 2 and 1,000 × *sqrt* 3. The diagonal across the rectangle is equivalent to 1,000 × *sqrt* 5. The reader will recall that values of *sqrt* 2, *sqrt* 3, and *sqrt* 5 are found many times over in the D&M Pyramid of Cydonia.

Another point about Giza that emerges from studying Legon's work (which was undertaken without any knowledge of Cydonia) is that the placement of the pyramid of Menkaure is seemingly defined by the Cydonian tetrahedral constant *t*.[19] The northwest corner of the pyramid of Menkaure is positioned on a line subtending 19.48 degrees from due south of the adjacent (southwestern) corner of the neighboring pyramid of Khafre. And the apex of the Menkaure pyramid is positioned exactly on a line subtending 19.52 degrees from southwest viewed from the same position.

GATEWAYS

If there are artificial pyramids on Mars filled with *pi, phi, e,* and *t* values and there are artificial pyramids on Earth filled with *pi, phi, e,* and *t* values, then the explanation must logically lie in one of four hypotheses:

1. There is no connection between the pyramids of Earth and the pyramids of Mars. The similarities are all coincidences.
2. An ancient Martian civilization that built pyramids came to Earth and taught the art of pyramid building to humans.
3. An ancient human civilization that built pyramids went to Mars and taught the art of pyramid building to Martians.
4. An ancient nonhuman civilization that built pyramids came from somewhere outside the solar system and left its mark on both Mars and Earth.

Of all these hypotheses we suggest that the first—coincidence—is the least likely to be correct. Common sense insists that *if* the pyramids of Mars are artificial, then there *must* be some connection with terrestrial pyramids.

More than 4,000 years ago, the pyramids of Giza were viewed by the ancient Egyptians as a gateway to the stars. The pyramids of Teotihuacan served exactly the same function for the ancient Mexicans. In both places men were believed to have been transformed into gods. In both places there were astronomical myths of great suggestiveness and complexity. In both places the monuments were said to reflect the pattern of heavenly prototypes. And in both places, as we were to discover, ancient texts and traditions show a special interest in the planet Mars.

17

The Feathered Serpent, the Fire-Bird, and the Stone

HUGH Harleston's calculations of the measures of the mysterious Mexican city of Teotihuacan eventually led him to the theory, which it is beyond our scope to explore here, that this site could inscribe a vast astronomical map—in which the distances between major structures stand in relation to the distances between the planets in the solar system.[1]

Harleston also developed a "quite unconventional" astronomical reading of an ancient Mexican myth concerning Xipe Xolotl, the twin brother of the high god Quetzalcoatl. The mythical bringer of civilization to Mexico at the beginning of the present epoch of the earth, Quetzalcoatl was often symbolized—notably at Teotihuacan itself—as a fiery feathered serpent (the name Quetzalcoatl means "feathered serpent"). Both Xipe Xolotl and Quetzalcoatl are also enigmatically spoken of in these myths as having been skinned—literally "flayed" alive (and, indeed, the flaying of sacrificial victims was a standard practice in ancient Mexico, particularly among the Aztecs, the last people to transmit the ancient myths before the coming of the Spaniards).

Harleston's reading sees the symbolism of Quetzalcoatl as referring, at one level, to

> a flayed planet—the twin to Mars—whose outer surface is conceived to have been deliberately "peeled-off like an orange. . . ." According to this

> reading, the damaged twin companion—Xipe Xolotl, the flayed Red God of the East, or Mars, retreated to a new position.[2]

This imaginative rendering does make us think.

As we have seen, Mars *is* technically a "flayed planet" with the hemisphere north of the line of dichotomy lying on average three kilometers lower than the southern hemisphere—which in its turn bears the scars of a cataclysmic bombardment. Could the myth of Xipe Xolotl be some garbled remembrance of such a catastrophe—involving the Red God of the East, Mars, having his skin flayed from his body by a "fiery serpent"? If so, then we are obliged to ask what real—as opposed to mythological—entity could fit the description of a fiery, feathered, or winged (and hence in some ways bird-like) serpent flying across the heavens with vivid plumes stretching out behind it.

Throughout history, and in all cultures, it may be significant that precisely such imagery has again and again been applied to comets. For example, Donati's comet of 1858, the "most splendid comet of the nineteenth century," was spontaneously described by eyewitnesses in the following terms:

> It had a head like a serpent, its body near the nucleus twisted and turned like a gigantic red serpent, and its tail, flashing like golden scales, spread over 40 million miles.[3]

We will see in part 4 that the nuclei of comets can be very large—up to several hundred kilometers in diameter—and can travel at speeds in excess of a quarter of a million kilometers per hour. Were such an object to strike a planet—Mars or Earth—it would certainly release sufficient impact energy to cause unimaginable devastation—perhaps even enough to "flay" its victim of its stony outer crust or "skin."

ASTRONOMICAL CYCLES

In Indian myth the god Vishnu lies sleeping on the cosmic ocean wrapped in the coils of the Naga serpent Ananda. From out of Vishnu's navel a lotus arises on which is seated the four-headed creator, Brahma. Brahma lives for one hundred Brahma years (vastly, infinitely longer than human years), on each day of which he opens and closes his eyes a thousand times. When he opens them a world comes into being; when he closes them a world comes to an end—a thousand worlds a day, millions of universes spawned and destroyed in his lifetime. When Brahma dies, the lotus closes, and withers. Then from Vishnu's navel a new lotus blooms, a new Brahma is born, and the process begins again.[4]

Each cycle of coming in and out of existence is itself subdivided into four stages, or epochs, called Yugas: the Krita Yuga (consisting of 1,728,000 human years), the Treta Yuga (1,296,000 human years), the Dvapara Yuga (864,000 human years), and finally the age in which we now find ourselves, the Kali Yuga (432,000 human years.)

Significantly, as Professor Hermann Jacobi has pointed out:

> The astronomical aspect of the yuga is that, in its commencement, sun, moon, and planets stood in conjunction in the initial point of the ecliptic, and returned to the same point at the end of an age. The popular belief on which this notion is based is older than Hindu astronomy.[5]

So the archaic marker for the end of an epoch is, in the final analysis, an astronomical one, an actual event in historical time, denominated in terms of the precession of the equinoxes. This is the cyclical process, described in the last chapter, which slowly shifts the zodiacal constellations against the background of which the sun rises on the spring equinox. (As the reader will recall, sun and stars were said to return to any arbitrarily defined starting point along the ecliptic—and the cycle begins again—once every 25,920 years.)

Not only in ancient India, but all around the world, it was understood that our present epoch of the earth is only one of a succession of such epochs—each with their own distinct and characteristic starting and ending points. Not only in ancient India, but all around the world, it was understood that the end of each cosmic epoch would be brought about by a cataclysm and followed by the birth of a new age.

PERIODIC DESTRUCTIONS

According to the Hopi of Arizona:

> The first world was destroyed, as a punishment for human misdemeanors, by an all-consuming fire that came from above and below. The second world ended when the terrestrial globe toppled from its axis and everything was covered with ice. The third world ended in a universal flood. The present world is the fourth. Its fate will depend on whether or not its inhabitants behave in accordance with the Creator's plans.[6]

In Aztec and Mayan mythology, we are living in the fifth epoch of creation, characterized as the "Fifth Sun." The fourth epoch was said to have been brought to an end by a great flood in which almost all men died ("There was water for fifty-two years and then the sky collapsed"). And it was prophesied that the fate of the fifth epoch—our own—is that it will end in a cataclysmic "movement of the earth" that will destroy civilization and perhaps even exterminate all traces of human life.[7] In the enormously sophisticated mathematical and calendrical system of the Maya, which we have explored at length in other works, the date of this coming cataclysm was foretold. That date is 4 Ahau 8 Kankin. When it is translated into the Gregorian calendar that we use today it becomes A.D. 23 December 2012.

Ancient Egypt also preserved complex beliefs concerning the cyclic creation and destruction of worlds. The little-known Edfu Building Texts,[8] for example, speak of a remote golden age, many thousands of years in the past, when the gods themselves lived on an island—the "Homeland of the Primeval Ones." The texts tell us this island was utterly destroyed in a terrible storm and flood caused by "a great serpent."[9] The majority of the "divine inhabitants" were drowned,[10] but the survivors of the cataclysm settled in Egypt where they became known as "the Builder Gods, who fashioned in the primeval time, the Lords of Light."[11] According to the Edfu Texts, it was these survivors who set out the foundations of all the future pyramids and temples of Egypt and who handed down the religion that would much later be practiced throughout the land under the semi-divine rule of the Pharaohs.

THE BENBEN OF HELIOPOLIS

The religious system practiced at the pyramids of Giza in Egypt was administered from the nearby sacred city of Heliopolis and had, as its central icon, a pyramidal stone called the Benben, which was said to have been made of *bja* metal (literally, "metal from heaven"). As we have argued at greater length elsewhere, there seems little doubt that this object, which was venerated in a special temple at Heliopolis called Het Benbennet—literally "Mansion of the Phoenix"—was a fragment of an iron meteorite.[12]

Essentially there are two sorts of meteorites: stone and iron. The iron, for obvious reasons, tend to be black and often larger than the stone variety, since they suffer little or no damage when they hit soft ground. Also, when entering the earth's atmosphere, some iron meteorites retain their direction of flight rather than roll about. These are called "oriented"—that is, they maintain their orientation as they fall, like an arrow or pointed artillery shell. As these oriented meteorites are heated during their fiery fall, their front part tends to melt and taper down. When found, therefore, they characteristically have the shape of a cone. Two good examples are the large conical—indeed almost pyramid-shaped—meteorites Williamette (to be seen in the American Museum of Natural History, New York) and Morito (presently on exhibit at the Danish Institute of Metallurgy).[13]

Many religious cults that venerated sacred meteorites existed in the ancient world. The cult at Delphi certainly had a meteoritic origin.[14] Pliny (A.D. 23–79) reported that a "stone which fell from heaven" was worshiped at Potidae.[15] The cult of meteorites was particularly rife in Phoenicia and Syria.[16] The sacred black stone of the Kaaba in Mecca is believed to be a meteorite.[17] And in ancient Phrygia (central Turkey) the great Mother of the Gods, Cybele, was represented at the temple of Pessinus by a black stone fallen from the sky.[18]

Sir E. A. Wallis Budge was the first scholar to suggest that the Benben stone of the ancient Egyptians must have belonged to this class of objects.[19] Subsequently another Egyptologist, J. P. Lauer, independently concluded that the Benben could only have been a meteorite.[20] Our own research has also convinced us of the very high probability that a large, oriented iron meteorite may have fallen near Giza some time in the first half of the third millennium B.C. From depictions of the Benben stone, it would seem that this meteorite was from 6 to 15 tons in mass, and the frightful spectacle of its fiery fall would have been impressive. The fall would have been presaged by loud detonations caused by shock waves, and even in daylight a fireball with a long plumed tail would have been visible from far away. Rushing to the spot where it landed, people would have seen that the "fire-bird" had disappeared, leaving only a black, pyramid-shaped *bja* object, or cosmic egg—the oriented iron meteorite.

FLIGHT OF THE PHOENIX

Intimately linked to the Benben in terms of symbolism and religious significance, and stemming from the common root word *ben,* was the *bennu* bird, the ancient Egyptian phoenix—the cult of which was also centered at Heliopolis. At widely separated cyclic intervals of many thousands of years, this creature was said to have

> fashioned a nest of aromatic boughs and spices, set it on fire and was consumed in the flames. From the pyre miraculously sprang a new Phoenix, which, after embalming its father's ashes in an egg of myrrh, flew with the ashes to Heliopolis where it deposited them in the altar of the sun-god Ra. A variant story made the dying Phoenix fly to Heliopolis and immolate itself in the altar fire, from which the young Phoenix then rose. . . . The Egyptians associated the Phoenix with immortality.[21]

Comparable in many ways to Quetzalcoatl, the fiery-winged (bird-like) serpent,[22] the defining qualities of the *bennu*/phoenix are therefore as follows:

1. It is a thing that flies.
2. It is a thing that returns, after long intervals.
3. It is a thing that is "consumed in flames."
4. It is in some way reborn, or renewed, on each return.
5. It is closely associated with the Benben meteorite—an iron "egg" fallen from heaven, which the ancient Egyptians are known to have kept in Het Benbennet, the "Mansion of the Phoenix" in Heliopolis.

CODE FOR A COMET?

It is often a mistake to give literal interpretations to the symbols of ancient religions. And we accept that the *bennu* and the Benben must be ranked among the most complex, subtle, and sophisticated symbols to be found anywhere in the ancient world. We have explored the spiritual implications of this symbolism elsewhere in our work.[23] But it is a characteristic of such powerful images as the phoenix and the stone that they can be employed at many different levels of meaning.

If we take the images literally and start looking around in the natural world for something that flies, which returns at cyclical intervals, which has the appearance of being "consumed in flames," which is mysteriously "renewed" on each occasion, and which is associated with meteorites, then there is really only one class of objects known to scientists today that could possibly fit the bill.

These objects, once again, are comets—the same objects symbolized in Mexican myths as fiery feathered or winged serpents, which we shall have the opportunity to investigate in part 4. Comets are responsible for several spectacular meteor showers that the earth encounters every year—showers consisting of relatively small scattered chunks of fragmenting parent comets, which continue to circulate in the same orbits as the showers. The family resemblance is obvious:

- Comets can truly be said to be associated with meteorites in much the same relationship as the "parent" *bennu*-phoenix to the "offspring" Benben stone that falls to earth.
- Comets, of course, "fly."
- Since comets are in orbit they also return to our skies at cyclical intervals—some as short as 3.3 years, in the case of Encke's comet, some longer than 4,000 years, in the case of comet Hale-Bopp, some even running into tens of thousands of years.
- Comets do literally undergo a process of renewal—indeed, rebirth—on each appearance in our skies. This is because comet nuclei are usually inert and utterly dark while traveling through deep space, producing no characteristic glowing coma and sparkling tail. However, as a comet approaches the Sun (and Earth), the solar rays cause volatile materials buried in its interior

to burst into boiling, seething activity, producing jets of gas—scientists call the process "outgassing"—and shedding millions of tons of exceptionally fine dust and debris to form the coma and the tail.

- Last but not least, outgassing comets do have the appearance of being consumed in flames—and the collision of any cometary fragment with Earth itself could lead to a gigantic, even worldwide, conflagration, *followed by a global flood,* as we shall see in part 4.

CLUES IN THE STARRY LANDSCAPE

The religion of the phoenix and the Benben, practiced at Heliopolis in the Pyramid Age—and for which the pyramids and the Great Sphinx of Giza were undoubtedly the central spiritual monuments—conveyed a distinctive system of teachings, which we have explored in several previous books.[24]

According to this religious system the afterlife journey of the soul is undertaken in a region of the sky known as the Duat—which has specific coordinates, demarcated on one side by the constellation Leo and on the other by the constellations Orion and Taurus. Through the middle of this skyscape, at the bottom of a wide, dark "valley," flows the celestial counterpart of the sacred river Nile—the stunning feature that we now call the Milky Way and that the ancient Egyptians knew as the "Winding Waterway."[25]

The gist of our previous work has been to show that it is not only the Milky Way that has a terrestrial twin in Egypt. The constellation Orion, represented by its three belt stars, is mirrored by the three pyramids of Giza.[26] The constellation Taurus, represented by two bright stars in the characteristic V of its horns is twinned with the two pyramids of Dashur.[27] And the constellation Leo has as its terrestrial counterpart the lion-bodied Sphinx of Giza.[28]

We saw in chapter 16 that precession alters the positions of all the stars in the sky according to a great cycle of 25,920 years—a cycle that proceeds at the rate of one degree every 72 years and that is most easily observable (although not within the short span of one human lifetime) as the precession of the equinoxes.

In *The Orion Mystery, Fingerprints of the Gods,* and *The Message of the Sphinx,* we have demonstrated with a substantial body of evidence that the pattern of stars that is duplicated on the ground at Giza in the form of the three pyramids and the Sphinx represents the disposition of the constellations Orion and Leo as they looked at the moment of sunrise on the spring equinox during the astronomical Age of Leo (the epoch in which the Sun was "housed" by Leo on the spring equinox).

Like all precessional ages, this was a 2,160-year period. It is generally calculated to have fallen between the Gregorian calendar dates of 10,970 B.C. and 8810 B.C.[29] In this epoch, and in no other, computer simulations of the effects of precession show that the three stars of Orion's belt—when viewed at dawn on the spring equinox—would have stood due south at the meridian in the pattern of the three pyramids on the ground, and that the Sun rose due east, in line with the gaze of the Sphinx, with the constellation Leo—the Sphinx's celestial counterpart—poised directly above it.[30]

There is geological evidence, which we will not repeat here, that the Sphinx may in fact date back as far as the eleventh millennium B.C.[31] But we do not dispute that the pyramids were built, or mostly built, during the *third* millennium B.C.—the date attributed to them by Egyptologists. Moreover, although we are satisfied that the ground plan of the Giza necropolis was conceived as an image of the equinoctial sky in the Age of Leo—10,970 B.C. to 8810 B.C.—we also note that the Great Pyramid has pronounced astronomical connections to the much later epoch of 2500 B.C. (the date at which Egyptologists believe it was built). These connections, which could not be more explicit, are the carefully angled shafts that emanate from the so-called King's Chamber and Queen's Chamber.[32] There are two shafts in each chamber, one of which points due north and the other due south. In the epoch of 2500 B.C.—and *only* in that epoch—precessional calculations show that all four of the shafts would have lined up like gunsights on the meridian transits of four stars that are known to have been of great significance to the ancient Egyptians:

> From the Queen's Chamber, the northern shaft is angled at 39 degrees and was aimed at the star Kochab (Beta Ursa Minor) in the constellation of the Little Bear—a star associated by the ancients with "cosmic regeneration" and the immortality of the soul. The southern shaft, which is angled at 39 degrees 30′, was aimed at the bright star Sirius (Alpha Canis

Major) in the constellation of the Great Dog. This star the ancients associated with the goddess Isis, cosmic mother of the kings of Egypt.[33]

From the King's Chamber, the northern shaft is angled at 32 degrees 28′ and was aimed at the ancient Pole star, Thuban (Alpha Draconis) in the constellation of the Dragon—associated by the pharaohs with notions of "cosmic pregnancy and gestation." The southern shaft, which is angled at 45 degrees 14′, was aimed at Al Nitak (Zeta Orionis), the brightest (and also the lowest) of the three stars of Orion's belt—which the ancient Egyptians identified with Osiris, their high god of resurrection and rebirth and the legendary bringer of civilization to the Nile Valley in a remote epoch referred to as Zep Tepi, the "First Time."[34]

A VAST AND EXTRAORDINARY STATEMENT

Because we can reconstruct the ancient skies over Giza with modern computers we can demonstrate the perfect alignments of the four shafts to the four stars circa 2500 B.C. What the same computers also show us is that these alignments were rare and fleeting, only valid for a century or so. After that the continuous gradual change effected in stellar declinations by the passage of time altered the positions at which the stars transited the meridian. It therefore seems inescapable—whatever their connections to the date of 10,500 B.C.—that the pyramids are also signaling an extremely close connection to the date of 2500 B.C.

Indeed, we are prepared to go further. It is our hypothesis that one of the multiple and complex functions of the monuments of the Giza necropolis may have been to make some sort of *statement* about two widely separated astrological ages—the Age of Leo, 10,970 B.C. to 8810 B.C. (in which falls the earlier date spelled out by the ground plan), and the Age of Taurus—that is, when Taurus housed the Sun on the spring equinox, generally put at 4490 B.C. to 2330 B.C. (in which falls the later date spelled out by the star shafts).

Only a statement of vast and extraordinary significance could have justified such a vast and extraordinary undertaking—for any rational analysis of the pyramids shows that they must have been built with enormous, almost unlimited resources, and the focused attention of the very best minds of the age over a sustained period of time. Indeed, their standards of precision are so high, and coupled with the use of such gigantic megaliths, that it is not certain they could be built today even with the best modern technology. Then, and now, they stand at the very edge of what is possible.

What were the ancients trying to say that they felt was worth such a superhuman effort?

GODS AND THEIR STARRY COUNTERPARTS

The pyramids and the Great Sphinx of Giza are uninscribed monuments that have never been proven to be just "tombs and tombs only," as Egyptologists like to tell us. Indeed, all that the monuments tell us about themselves—from their alignments, their shafts, and the presence within them of empty sarcophagi—is that their builders connected them to the stars, to the cyclical flow of time as measured by precession, and to ideas about death. The Heliopolitan religion that was practiced around them, however, has left us a gigantic legacy of texts, some inscribed on the walls of later pyramids (the so-called Pyramid Texts), which help to fill in the picture.

We have already encountered the Heliopolitan symbolism of the Benben stone and the *bennu*-phoenix. It is as well to remind ourselves also of some of the principal Heliopolitan gods and of their astronomical counterparts:

- Atum-Ra, the creator, the father of the Gods, identified with the Sun.[35]
- Osiris, the first divine pharaoh of Egypt, later transformed into the god of death and rebirth, associated with the constellation of Orion.[36]
- Isis, goddess of magic, sister and consort of Osiris, associated with the star Sirius.[37]
- Set, god of storms and chaos, violence and darkness, fire and brimstone, the murderer of Osiris and usurper of his kingdom, associated with the constellation of Taurus.[38]
- Horus, the revenging son of Osiris and Isis who defeats Set and restores his father's kingdom, associated with the constellation of

Leo, with the Sun when it is "in" the constellation of Leo, and also with a planet that sometimes passes between the paws of the constellation of Leo—the planet Mars, as we shall see.[39]

MESSAGE OF THE CATACLYSM

The Egyptian golden age over which Osiris was said to have presided is referred to in the Pyramid Texts as Zep Tepi—literally the "First Time." This word *tepi,* as we have shown in *The Message of the Sphinx,* refers to a new cycle of time ushered in symbolically by the appearance of the phoenix flying from the east, alighting in Heliopolis, and starting time with its cry. We are now beginning to wonder, however, if it is only a *symbolic* ushering in that was intended here. Or is it possible that the "phoenix" with its fiery, meteoritic associations could in fact *be* a comet, as we suggested above. Perhaps a comet that was seen to return to the skies over Egypt at cyclical intervals, on each occasion capsizing the old order of the world and ushering in the new?

We suspect, and have argued at length in our works,[40] that the story of the golden age of Osiris may have historical foundations in a lost prehistoric civilization—highly advanced both scientifically and spiritually—that was obliterated more than 12,000 years ago in the huge global cataclysm that shook the earth at the end of the last Ice Age.

No scholar today doubts that such a cataclysm occurred—more than 70 percent of all animal species were rendered extinct—but the more interesting and still unresolved issue is, what caused it?

As we will show in part 4, evidence has been steadily accumulating during the past decade that does indeed link the whole mystery to a fragmenting giant comet, trapped in a cyclical Earth-approaching orbit, that was responsible for massive impacts in the eleventh millennium B.C. and in the ninth millennium B.C.—the exact span of the Age of Leo—and for a later episode of bombardment in the third millennium B.C., toward the end of the Age of Taurus, at around the time that the Giza pyramids were built.

Is it simply some sort of bizarre coincidence that one level of the enormously sophisticated multi-layered message that ancient Egypt has passed down to us could, with perfect legitimacy, be read as follows:

Bennu/Phoenix	=	Large, Earth-approaching comet
Benben/Stone	=	Meteoritic debris of the same comet
Ground plan of the pyramids and the Great Sphinx of Giza	=	Signpost written in the universal language of precessional astronomy stating that the comet (phoenix) visited Earth in the Age of Leo, the mythical golden age called *Zep Tepi* in the Egyptian calendar, 10,970–8810 B.C.
Star shafts of Great Pyramid	=	Signpost written in the universal language of precessional astronomy remarking on the return of the phoenix to the close vicinity of Earth during the Age of Taurus, 4490–2330 B.C.

DANGER FROM TAURUS?

A curious matrix of mythology surrounds the essential symbolism and architecture in which the phoenix story unfolds.

As we have seen:

Osiris	=	Orion
Isis	=	Sirius
Set	=	Taurus
Horus	=	Leo

We also know that in the Heliopolitan myths Set killed Osiris and usurped his kingdom (interestingly enough, with the help of 72 co-conspirators,[41] and 72 is the key number in the precessional code, outlined in chapter 16). The myths further state that Isis/Sirius used her magic to restore Osiris briefly to physical life so that she could copulate with him and receive his "seed." He was then translated to the heavens where he became judge of the dead and god of rebirth. Meanwhile, as we noted earlier, the fruit of his union with Isis was Horus, who in due course was destined to grow to manhood, overthrow Set, and restore his father's kingdom.

New life, the myth seems to be saying, comes from the death of the old—literally the dead body of the old god. In a sense, the image of Osiris-Horus is the same as that of the phoenix. Just as the immolation of the phoenix ends the previous world age, so the death of Osiris ends Zep Tepi and leads, ultimately, to the reign of the pharaohs.

But we know that all the principal players in the drama have stellar counterparts, so it is also worth considering the myth at a more literal, astronomical level:

1. The villain of the piece is Set, who murdered Osiris and ended the golden age.
2. Set is strongly identified with the constellation Taurus.
3. This therefore implies that Taurus must have been seen by the ancient Egyptians as a source of danger, chaos, and destruction.

RED PLANET, RED SPHINX

The Egyptian name of the Sphinx was Horakhti, "Horus of the Horizon," the manifestation of the sun god at the moment of rising. We have shown in *The Message of the Sphinx* that the very same name, Horakhti, was applied to the constellation Leo.[42] In addition, as the eminent Egyptologist Sir E. A. Wallis Budge points out, the name Horus—originally Heru—conveys the meaning "face"; thus the name of the Sphinx could mean "Face of the Horizon"—referring to the face of the solar disk.[43]

Inevitably some of the AOC researchers have made much of this to connect it to the Face on Mars—something for which there would be no justification were it not for a series of peculiar clues that seem to point in the opposite direction:

1. As Richard Hoagland was the first to realize, the city of Cairo, on the southern edge of which the Giza necropolis stands, got its

present name in the tenth century A.D. from invading Arabs who inexplicably decided to call it El-Kahira, meaning "Mars."[44]

2. The name the ancient Egyptians gave to the planet Mars was Hor Dshr—literally "Horus the Red."[45]
3. In inscriptions found in certain tombs in upper Egypt, Mars is also referred to as "His name is Horakhti" and as "the eastern star."[46] Since the gaze of the Sphinx is oriented precisely due east, and since the Sphinx was likewise called Horakhti, as we have seen, we may just as well say that the name of the Sphinx is Mars.
4. Along with all the other planets, and the Sun itself, Mars appears to travel in an endless cycle through all twelve constellations of the zodiac. This means that it will, at intervals, be seen to pass through the constellation of Leo—to be "in" Leo or "housed" by Leo, in astrological parlance.
5. For a long period of its history the Sphinx was painted red.[47]
6. Since the Sphinx is a composite creature with the head of a man and a body of a lion, we also note in passing that ancient Hindu myths depict the planet Mars as Nr-Simha, the "Man-Lion."[48]

What all these clues suggest to us, at the very least, is that the ancients must have seen a clear and direct association between the Red Planet and the Sphinx. Moreover, since the astronomy of the Sphinx is so precisely set to the rising of the constellation Leo at the spring equinox in the epoch 10,970 B.C. to 8810 B.C., we suspect that part of the message may be to consider events that could have visibly affected both the planet Mars and Earth during this epoch—that is, the astronomical Age of Leo. There is also a strong hint in the surrounding mythology to suggest that such events, whatever they may be, are likely to turn out to be connected in some way to Taurus, the Bull of the Sky—the constellation of Set the destroyer.

The classical Greeks, who sat at the feet of the ancient Egyptians and learned everything they knew from them, renamed Set as Typhon and depicted him as a terrifying supernatural monster whose

> head touched the stars, his vast wings darkened the Sun, fire flashed from his eyes, and flaming rocks hurtled from his mouth. When he came rushing toward Olympus the gods fled in terror to Egypt.[49]

Likewise the Roman historian Pliny (A.D. 23–79) writes of a remote epoch during which "a terrible comet" given the name Typhon was seen by the people of Egypt:

> It had a fiery appearance and was twisted like a coil and it was grim to behold. It was not really a star so much as what might be called a ball of fire.[50]

We wonder whether it is possible, in their architecture and in their myths, that what the ancients were trying to pass down to us might have included a package of lifesaving data:

- Their recollections of the breathtaking returns to the inner solar system of a fiery and spectacular periodic comet
- Specific information about this comet's previous dangerously close approaches to Earth
- Specific information about at least one cataclysmic approach the comet made to Mars that "flayed" the Red Planet of its skin
- Specific information about if and when the threat will return to menace us, and perhaps even information about the the direction from which it will come (the direction of the constellation of Taurus?).

Today there is no fear of comets. Indeed, we hardly ever even stop to look at the skies. But to the ancients they were terrible instruments of doom and destruction "importing change of times and states"[51] and shaking "pestilence and war" from their "horrid hair."[52] We will see in part 4 that this ancient reputation may be nothing less than the truth, and that comets may indeed be agents in the destruction and rebirth of worlds.

PART FOUR

The Darkness and the Light

18

The Moon in June

ON the evening on 25 June 1178, five friends were sitting out after dark on the outskirts of the English cathedral city of Canterbury, chatting and enjoying the summer air.[1] The sky was cloudless and a bright new moon was rising with its horns tilted toward the east. Then suddenly:

> The upper horn split in two. From the midpoint of the division a flaming torch sprang up, spewing out, over a considerable distance, fire, hot coals and sparks. Meanwhile the body of the Moon which was below writhed, as if it were in anxiety, and to put it in the words of those who reported it to me and saw it with their own eyes, the Moon throbbed like a wounded snake. Afterwards it resumed its proper state. This phenomenon was repeated a dozen times or more, the flame assuming various twisting shapes at random and then returning to normal. Then, after these transformations, the Moon from horn to horn, that is, along its whole length, took on a blackish appearance. The present writer was given this report by men who saw it with their own eyes and are prepared to stake their honour on an oath that they have made no addition or falsification in the above narrative.[2]

The writer is the twelfth-century monk Gervase of Canterbury, whose *Chronicle* is highly regarded as a work of history. Because of his renowned accuracy scholars generally agree that "Gervase's record of the 'Canterbury Event' must be taken seriously."[3]

Yet if it is a true report, then what is the strange phenomenon it describes?

In 1976 the American astronomer Jack Hartung offered an answer that most scientists now accept. He deduced that Gervase's eyewitnesses saw the cataclysmic effects of a collision between the Moon and some large object flying through space—such as a comet or an asteroid. He further reasoned that if he was correct then there ought to be an impact crater of suitable shape and size at an appropriate lunar latitude. Reckoning on the basis of the Gervase report, Hartung calculated that such a geologically recent crater would be

> at least 7 miles in diameter, possess bright rays extending from it at least 70 miles and lie between 30 degrees and 60 degrees north and 75 degrees and 105 degrees east.[4]

Named after an Italian heretic (burned at the stake in 1600 for professing the existence of inhabited planets other than Earth), the crater Giordano Bruno perfectly fits Hartung's bill. It has a radius of 13 miles and the telltale bright rays of a recent cataclysmic impact.[5] Moreover, although it lies almost 15 degrees into the dark side of the Moon, the astronomers Odile Calame and Derral Mulholland have demonstrated that the ejecta from the impact would have been hurled such distances that "the event would not only have been visible but sufficiently apocalyptic to have justified the description given in the Canterbury Chronicle."[6]

Calame and Mulholland's work provides additional confirmation that the Moon has indeed suffered a major impact at some time during the past millennium. In research conducted between 1973 and 1976 they used the 107-inch reflector telescope at the McDonald Observatory in West Texas to direct more than 2,000 laser beams at a series of mirrors left behind on the Moon by *Apollo* astronauts. The beams allowed extremely precise measurements to be made and revealed "a 15-meter oscillation of the lunar surface about its polar axis, with a period of about three years."[7] As the American cometary astronomer David Levy puts it, the Moon is behaving just "like a huge bell vibrating after it has been clanged."[8] Two leading British astronomers, Victor Clube of Oxford University and his colleague Bill Napier of the Royal Armagh Observatory, point out that such a mode of vibration "dies out over 20,000 years or so" and confirm that "the result can only be explained by a recent large impact, whose magnitude was about that required to form the Bruno crater."[9]

The crater was made by an object estimated by scientists to have been around two kilometers in diameter, which exploded on impact with the energy of 100,000 megatons of TNT—that is, 100,000 *million* tons of TNT, roughly equivalent to ten times the explosive power of all the nuclear weapons presently stockpiled on earth (although, of course, without the radioactive fallout).[10] By contrast, the atomic bomb that obliterated the Japanese city of Hiroshima in 1945 had a payload of 13 kilotons (13 *thousand* tons of TNT) and the largest individual nuclear weapons in existence today have yields rated at approximately 50 megatons.[11]

At 100,000 megatons, it is easy to see why some historians believe that the Canterbury Event could have wiped out human civilization on 25 June 1178 if it had occurred on Earth rather than on the Moon.[12]

TUNGUSKA

Eight hundred and thirty years later, on 30 June 1908, a much smaller object did hit Earth—with devastating consequences. This was the event that flattened more than 2,000 square kilometers of forest in the Siberian wilderness region of Tunguska. It was an airburst, not a land impact, involving the explosive fragmentation of a bolide with an estimated diameter of 70 meters at an altitude of about 6,000 meters.[13]

We described some aspects of the Tunguska Event in chapter 4. Its effects were dramatic. The bolide, descending as a huge fireball, was said to be brighter than the sun and was visible at a distance of more than 1,000 kilometers from the blast zone.[14] It is estimated to have been traveling at a speed of 30 kilometers per second and was said by those who saw its passage to have emitted a series of intense thunderclaps. When it exploded it did so with a "stupendous bang" that could be heard more than 1,000 kilometers away.[15]

The firestorm rapidly fell from the atmosphere to the ground, but as soon as contact was made a raging "column of fire" leapt up again from ground to sky. Several eyewitness accounts indicate that this fiery pillar may have been as much as 1,500 meters wide and 20 kilometers high, and that it was visible to observers as far away as 400 kilometers.[16]

> The whole northern sky appeared to be covered with fire [reported a farmer who had been at the Vanavara trading center just 60 kilometers from the blast zone]. I felt a great heat as if my shirt had caught fire. Afterward it became dark and at the same time I felt an explosion that threw me from the porch. . . . I lost consciousness.[17]

Another farmer, 200 kilometers from the blast zone recalled:

> When I sat down to my breakfast beside my plough, I heard sudden bangs as if from gunfire. My horse fell to its knees. From the north side above the forest a flame shot up. Then I saw that the fir forest had been bent over by the wind and I thought of a hurricane.[18]

At a distance of 400 kilometers the tremors set off by the Tunguska explosion were so intense that the Trans-Siberian Railroad had to be halted for fear of derailment.[19] There was also a devastating shock wave that mowed down the dense forests of the region, "snapping off meter-diameter trees like matchsticks"[20] and convincing some villagers that "the end of the world was approaching."[21] The impact energy of the blast was in the range of 10 to 30 megatons of TNT—at least seven hundred times more powerful than the Hiroshima bomb.[22] Little wonder, therefore, that as far away as Western Europe people reported "White Nights" for several evenings after the 30 June Tunguska explosion and were "able to read newspapers from the sky glow, even after midnight."[23]

The entire event, it must be remembered, was caused by an object 70 meters in diameter—that is, with a "footprint" about the size of a city block—rather tiny by cosmic standards. Because the explosion took place in a remote part of the world little attention was paid to it; indeed, it was not until 1927 that the first scientific expedition reached the site.[24] The expedition was led by the Soviet astronomer Leonard Kulik, who quickly realized from the extent of the devastation that if the same bolide had disintegrated in the skies above central Belgium "no creature would have been left alive in that country."[25] It is therefore sobering to recall that if the Tunguska object had collided with Earth *just three hours later than it did*—say at 10:00 in the morning instead of at 7:00 A.M.—it would not have laid waste an empty part of Siberia but would have exploded over the city of Moscow.[26]

At the very least we can say that such an accident would have changed the course of world history.

BOULDERS

The laser reflectors that Calame and Mulholland used in their research were not the only instruments that NASA's *Apollo* astronauts left on the Moon. Seismometers were also positioned at a variety of locations on the lunar surface to gather evidence of cosmic bombardments and transmit the data back to Earth.

From 1969 to 1974 nothing sensational happened. Then, over five consecutive days from 22 through 26 June 1975, the seismometers all burst into life in unison to record a roller-coaster event. The Moon had run into a swarm of boulder-sized meteoroids weighing about a ton each.[27] It received a sudden, unmerciful pounding—as many impacts in this five-day period as it had suffered in all of the previous five years.[28]

DEVASTATING EFFECTS

Along with the planets and their moons, vast quantities of rock, ice, and iron circulate within the solar system at breathtakingly high speeds, pursuing a tangled cat's cradle of chaotic and constantly changing orbits. Again and again fragments of this cosmic rubble intersect the orbits of the inner planets, notably Mars and the Earth-Moon system—sometimes with effects so devastating that any form of civilization unfortunate enough to be caught up in such a collision would certainly be wiped out. The final word has yet to be said on the true life story of Mars, but we know for certain that there have been a number of cosmic impacts that have come very close to obliterating not just "civilization" on Earth but *all* of this planet's animal and plant life.

IMPACTS AND CRUSTAL DISPLACEMENTS

Earth is thought to be 4.5 billion years old and has been a home to life—initially in the simplest forms—for perhaps 3.9 billion years. The oldest prokaryotic fossils date back about 3.7 billion years; the oldest eukaryotic fossils almost 2 billion years; and the oldest animal fossils about 800 million years.[29] Sometime between 550 million years ago and 530 million years ago our planet was overtaken by an immense cataclysm of unknown origin. Writing in *Science* on 25 July 1997, a group of researchers at the California Institute of Technology report that one of the terrible consequences of this event was a slippage of Earth's rigid outer crust around its inner layers.[30] The end result was "a 90-degree change in the direction of Earth's spin axis relative to the continents," commented Dr. Joseph Kirschvink, professor of geobiology at Cal Tech:

> Regions that were previously at the North and South Poles were relocated to the equator, and two antipodal points near the equator became the new poles. . . . The geophysical evidence that we've collected from rocks deposited before, during, and after this event demonstrates that all the major continents experienced a burst of motion during the same interval of time.[31]

The Cal Tech researchers insist that the event they are describing is to be distinguished entirely from "plate tectonics," an internal geological process of Earth that very slowly and gradually causes continental landmasses to drift apart or move together at a rate of no more than centimeters per year. What their evidence points to is a titanic rotation of the entire crust of Earth *in one piece* and at a cataclysmically fast rate. According to Kirschvink: "The rates . . . were really off the scale. On top of that everything [seems to have been] going the same direction."

We noted in chapter 4 that there is evidence of a major one-piece slippage of the crust of the planet Mars. No evidence has yet been offered as to how or why such a slippage could have occurred. Nevertheless, as the astronomer Peter Schultz has demonstrated, "Typical mantled and layered polar deposits have been found 180 degrees apart at the equator, i.e., in positions antipodal to one another—as would be expected with former poles."[32]

Two years before the publication of the Cal Tech article in *Science,* we reported in *Fingerprints of the Gods* on the recent work of Rand and Rose Flem-Ath in Canada, and the earlier work of Professor Charles Hapgood and Albert Einstein in the United States, which suggests that cataclysmic crustal displacements may have occurred on Earth, perhaps even as recently as the end of the last Ice Age.[33] Despite Einstein's prestigious support, this theory was ridiculed by orthodox geologists when Hapgood first proposed it in the 1950s and received a further dose of scholarly abuse when the Flem-Aths promulgated it again in their 1995 book *When The Sky Fell.*[34]

The essence of the orthodox "refutation" or "debunking" is that there is no known mechanism sufficiently powerful to set off crustal displacements and that such events are therefore "geophysical impossibilities." In this way intriguing pieces of evidence marshalled by crustal-displacement theorists have been repeatedly swept under the carpet. Even if an adequate mechanism has not yet been identified, however, the latest discoveries must surely shake the orthodox consensus. For what the Cal Tech researchers are talking about—this time under the banner of peer-reviewed respectability that *Science* represents—is nothing more nor less than a fully fledged displacement of the Earth's crust that could not have failed to have cataclysmic consequences.

It should therefore come as no surprise to learn that the extinction of an estimated 80 percent of all genera of life occurred at this time.[35] With almost miraculous speed, life then bounced back and the extinction was followed by

> a profound diversification that saw the first appearance in the fossil record of virtually all animal phyla living today. With relative evolutionary rates of more than 20 times normal, nothing like it has occurred since.[36]

This was the so-called Cambrian explosion, and it was indeed the greatest diversification and expansion of life that the earth had ever seen. Since then scientists believe that at least five further great extinctions—and about a dozen smaller ones—have occurred.[37] Evidence is growing that *all* these extinctions, as well as the gigantic crustal displacement that preceded the Cambrian explosion, may have been sparked by high-speed collisions with massive chunks of cosmic rubble on Earth-crossing orbits.[38] If they were to release sufficient impact energy, such collisions might theoretically provide the missing mechanism that scientists have been looking for that could set the crusts of entire planets in motion. One might even imagine a scenario for Earth in which all major impacts result in extinctions, but a sufficient energy threshold has to be crossed—or other conditions fulfilled—before an impact can trigger a crustal displacement.

IMPACTS AND EXTINCTIONS

One of Earth's five big extinctions took place at the juncture of the Permian and Triassic periods around 245 million years ago. Under mysterious circumstances 96 percent of all oceanic species and 90 percent of all land-dwelling species were wiped out at a stroke.[39] The radio astronomer Gerrit Verschuur, now professor of physics at the University of Memphis in Tennessee, comments:

> No localized flicker of nature can account for the sudden demise of so many species at the same time. It required a global phenomenon of staggering proportions. . . . Life on Earth very nearly came to an end. Words can barely begin to describe the enormity of such a catastrophe.[40]

Evidence has been presented linking this extinction with an impact—although geologists are be no means unanimous on the matter.[41] By contrast there is certainty regarding the later great extinction that took place 65 million years ago at the Cretaceous-Tertiary (K/T) boundary. Following breakthrough discoveries in the 1970s and 1980s,[42] all scientists today accept that this event was caused by a gigantic object from space—*an object at least 10 kilometers in diameter*—that smashed into the northern tip of the Yucatan peninsula at a speed of approximately 30 kilometers per second.[43] The resulting crater, now deeply buried beneath millions of years of accumulated sedimentation, has a diameter of almost 200 kilometers. It was first identified on gravitational maps made by surveyors looking for oil and subsequently confirmed by radioactive dating to be 65 million years old.[44]

As we noted in chapter 4, this K/T Boundary Event caused the extinction of the dinosaurs. It is also estimated to have killed off 50 percent of all other genera; 75 percent of species; and a staggering 99.99 percent of all individual animals then living on Earth.[45]

A GLOBAL CATACLYSM

The sequence of events and what exactly happened to our planet 65 million years ago has been reconstructed by scientists (who are generally of the opinion that the K/T object must have been a comet). According to the geologist Walter Alvarez:

> About 95 percent of the atmosphere lies below an altitude of 30 km, so depending on the angle at which the impactor approached the surface, it would have taken only a second or two to penetrate. The air in front of the comet, unable to get out of the way, was violently compressed, generating one of the most colossal sonic booms ever heard on this planet. Compression heated the air almost instantaneously until it reached a temperature of four or five times that of the Sun, generating a searing flash of light during that one-second traverse of the atmosphere.
>
> At the instant of contact with the Earth's surface, where the Yucatan peninsula now lies, two shock waves were triggered. One shock wave ploughed forward into the bedrock, passing through a three-kilometer-thick layer of limestone near the surface, and on into the granitic crust beneath. . . . Meanwhile a second shock wave flashed back into the onrushing comet.[46]

Gerrit Verschuur takes up the story:

> Within an hour of impact the rumble of the earth is heard around the world and earthquakes toss everything into the air. With magnitude 12 to 13 on the Richter scale the earthquake mangles solid rock as the ground buckles. All around the planet the seismic shock rumbles. As it travels its energy begins to focus so that at the antipodes it gathers and the planet's surface buckles and heaves 20 meters. . . . Eight hundred kilometers from the impact a tsunami more than a kilometer high washes over the North American continent to create ripples in the land that will be preserved and etched in geological strata for 65 million years to come. . . . A hundred meters of deposits dragged from the bottom of the sea cover the islands and the coastal regions of the mainland, and boulders the size of automobiles land 500 kilometers from the impact in a country later to be called Belize.[47]

Despite the tidal waves there is evidence that a global firestorm must have raged for several days after the K/T impact until it finally burned itself out. Scientists report the discovery of "a pervasive soot and charcoal layer . . . which indicates that upward of 90 percent of the biomass was incinerated at that time in global wildfires."[48]

Soon the world fell into a sort of "nuclear winter" as dust and smoke hefted up into the atmosphere by the impact and by the fires blotted out the light of the sun for several months.[49] Alvarez is of the opinion that "the land became so dark that you could not have seen your hand in front of your face."[50] A long period of freezing shadowy gloom then followed, during which many of the animal species that had survived the initial effects of the impact would have perished from cold, hunger, and exposure. Photosynthesis was suppressed and all over the earth the food chain was interrupted.

UNSEEN DANGERS

The explosive energy of the K/T object has been estimated at 100 million megatons of TNT—that is, about 1,000 times greater than the object that made the 13-mile Giordano Bruno crater on the Moon in 1178.[51] Yet an impact of either magnitude would be a civilization-destroying event—and perhaps the end of all mankind—if it were to afflict Earth today.[52] Indeed, as we saw in part 1, sufficiently large impacts such as those that struck Mars at some point during its history are capable, under certain circumstances, of sterilizing an entire planet.

Ours is a resourceful species that has survived through its ability to adapt to threats and to anticipate dangers. Is it not obvious from the terrible fate suffered by Mars, and from the evidence of past impacts on Earth and on the Moon, that we should pay attention to the possibility that unseen dangers may be lurking in the dark reaches of space among the planets of the solar system?

19

Signs in the Sky

In 1990, David Morrison, an astronomer at the NASA Ames Research Center, observed wryly that "there are more people working in one fast-food restaurant than there are professionals scanning the sky for asteroids."[1] This is no longer quite true today. Public funding for such work is still so miniscule as to be almost laughable—indeed, the grand total of all contributions from all governments worldwide rarely exceeded a million dollars a year from 1990 until the end of 1997.[2] Nevertheless Spacewatch programs that scan the sky for asteroids have been established in a number of countries, relying heavily upon concerned astronomers who are prepared to volunteer their time.[3]

At Kitt Peak National Observatory in Tucson, Arizona, which does receive some of NASA's limited Spacewatch funding, a team of astronomers is involved in a systematic long-term search for near-Earth asteroids using a 90-centimeter telescope and a CCD camera. The program is reported to have discovered "an average of two or three near-Earth objects each month, the smallest only 6 meters across."[4]

Related Spacewatch investigations include the Near-Earth-Asteroid Tracking Program of the U.S. Air Force observatory in Hawaii; the Planet-Crossing Asteroid Survey at Palomar Mountain in California; the asteroid search program of the Côte d'Azur observatory in Southern France; and the Anglo-Australian Near-Earth Asteroid Survey (which was terminated due to lack of funds in 1996).[5]

Will more resources be forthcoming for such programs in the future?

This is an area in which policymakers tend to be long on promises and short on action. But we do take it as a sign of an important change of heart—albeit one that has predictably not yet resulted in any more money—that the U.S. House of Representatives wrote the following clause into the NASA Authorization Bill of 20 July 1994:

> To the extent practicable, the National Aeronautics and Space Administration, in coordination with the Department of Defense and the space agencies of other countries, shall identify and catalogue within 10 years the orbital characteristics of all comets and asteroids that are greater than 1 kilometer in diameter and are in an orbit around the Sun that crosses the orbit of Earth.[6]

Why greater than one kilometer in diameter? The reason is a generally held belief that human civilization could survive a collision with a half-kilometer object and might not survive a collision with an object more than one kilometer wide. But what about a *swarm* of half-kilometer objects—or of quarter-kilometer objects, for that matter—or even a swarm of Tunguska-sized bolides penetrating Earth's atmosphere repeatedly, over hundreds of different locations, for a period of a week or two? Would that be survivable? And could it happen?

CRATERS

During the last two centuries astronomers have learned a great deal about the solar system and about near-Earth space—and nothing that they have discovered is reassuring. On the contrary, as our planet orbits the sun at a steady velocity of almost 110,000 kilometers per hour, we now know that it passes repeatedly through "lumpy" streams of cosmic debris. Most of the rubble takes the form of tiny meteors that burn up harmlessly in our atmosphere in the form of shooting stars. But there are also larger objects that explode in the sky and even more massive objects that make it to the ground. As we have seen, Earth during its long history has several times collided with such objects from space. Moreover, it is clear that the Tunguska and K/T events reported in the preceding chapter are by no means isolated incidents. According to the astronomer Sir Fred Hoyle, Earth could well have suffered more than *130,000 major impacts* over the last billion years.[7]

One worrying feature is that many impacts appear to have involved groups of objects rather than just individual projectiles. We have mentioned the prospect of "swarms of Tunguskas"—in itself a nightmarish possibility, as we shall see—but it is now clear from the geological record that the 10-kilometer object that caused the K/T event was also part of a swarm. At least a dozen other craters with dates indistinguishable from the K/T event have been found. These include the totally buried 35-kilometer "Manson structure" in Iowa state.[8]

Because the earth's surface is dynamic and subjected to continuous erosional and depositional forces, even the largest craters can and do disappear in matters of millions of years. In addition, because water covers seven-tenths of the surface of this planet, simple logic suggests that the majority of impacts must take place in the oceans—where they leave fewer long-term traces than impacts on land. Another important factor is that it is only since the late 1920s that impact craters have been recognized for what they are, having previously been wrongly attributed to volcanism, so this is a relatively new area of study.[9] Nevertheless more than 140 major craters have now been firmly identified, distributed all around the earth, and about five more are found every year.[10] Although some are as much as 200 million years old, surprising numbers of them are recent.[11]

Interesting discoveries include a chain of craters in South America made by a swarm of small iron meteorites. The meteorites appear to have entered the atmosphere at a shallow angle, only surviving because of their iron (as opposed to stony) constitution and then impacting Earth along a narrow 18-kilometer track in the region of Campo del Cielo, Argentina.

> Individual meteorites of different sizes were well sorted by sequence of mass along the track, evidently by aerodynamic (drag) forces. Disruption of the parent body occurred at an altitude of several kilometers. Radiocarbon dating of charcoal from one of the craters suggests that the event occurred well within the time of human occupancy in South America, about 2900 B.C.[12]

A second crater chain thought to be "no more than a few thousand years old" lies in the heart of the Argentinian pampas and was first spotted by an air force pilot flying overhead in 1989.[13] It is 30 kilometers from end to end. Its craters are not circular, as is the case with vertical impacts, but elongated—the three largest are each four kilometers long by one kilometer wide. Numerous smaller craters "were evidently made by fragments hurled downrange."[14]

More than 10 percent of Earth's craters larger than half a kilometer across have at least one companion crater nearby,[15] and three of the largest impact structures on Earth are conspicuously paired with smaller ones: the Steinheim and Reis craters in Germany (46 kilometers in diameter and 24 kilometers in diameter, respectively), which are both 15 million years old; the Kamensk and Gusev craters in Russia, both 65 million years old; and the twin Clearwater Lakes in Canada in northern Quebec, east of Hudson's Bay, which are 290 million years old.[16]

Lake Manicougan in Canada is an impact crater 60 kilometers in diameter.[17] The Sudbury structure in Ontario, containing one of the world's largest deposits of nickel and other valuable metals, is now recognized as "a tectonically distorted impact crater that was initially about 140 kilometers in diameter."[18] The 100-kilometer-diameter Vredfort Dome in South Africa is an impact structure.[19]

Astronomer Duncan Steel, head of Spaceguard Australia and founder of the Anglo-Australian Near-Earth Asteroid survey, estimates:

> We have yet to discover more than 1 percent of the impact structures on Earth. . . . Hundreds of craters are undoubtedly still hidden beneath the forest canopy of the Amazon basin, the tundra of the Arctic regions . . . the shifting sands of northern Africa and Arabia . . . [and] the 70 percent of Earth covered by water. . . . So far only one submarine crater has been found, the 60-kilometer-wide, 50-million-year-old Montagnais structure in the coastal waters of Nova Scotia.[20]

Yet the inventory of Earth's impact craters continues to grow. When set alongside the horrific scars of Mars and the pockmarked face of the Moon it should remind us that the solar system is and always has been a hazardous place—hazardous to all planets and all life in all past epochs and, obviously, still hazardous today.

ASCLEPIUS AND HERMES

In 1989 an asteroid with an estimated diameter of half a kilometer crossed Earth's path. "Earth had been at that point in space only six hours earlier," a House of Representatives committee report noted.

> Had it struck Earth it would have caused a disaster unprecedented in human history. The energy released would have been equivalent to more than 1,000 one-megaton bombs.[21]

With the dimensions and stored kinetic energy of "a giant aircraft carrier traveling at a speed of 42,000 miles per hour,"[22] this object was not detected by any astronomer until three weeks *after* it had thundered past us.[23] Now catalogued as 4581 Asclepius, it came, at its closest, to within 650,000 kilometers of Earth.[24]

This was a new record close passage—though we will see that it did not stand for long. The previous closest passage had been registered in 1937 by Hermes, a somewhat larger asteroid (estimates of its diameter range between one and two kilometers).[25] On the night before Halloween it approached Earth at alarming speed, "moving at up to 5 degrees an hour and completely crossing the sky in nine days."[26] The effect, according to an astronomer at the time, was "much like that obtained by standing near the railroad track when the evening express roars past."[27]

After staging this breathtaking flyby, Hermes vanished into the darkness of space and has never been seen again—an unsatisfactory state of affairs since past close passages make future close passages more likely.[28] Hermes is therefore an object to be watched. We can be sure that it is still lurking in the solar system and there is a fair chance that it has crossed the track of our planet's orbit more than once since 1937 but has simply not been spotted.[29] Asteroids of this size are extremely easy to miss in telescopic surveys and, as we shall see, astronomers believe that several thousand of them may be circulating in our immediate neighborhood.

INCOMING ASTEROIDS

On Sunday, 19 May 1996, and again less than a week later on 25 May 1996, Earth was approached by two potentially apocalyptic asteroids. The first—catalogued as 1996 JA—zoomed past at a distance of about half a million kilometers and at an estimated speed of 60,000 kilometers per hour. Astronomers were able to give us only four days' advance notice of its arrival on our cosmic front porch. The second, asteroid JG, was more than a kilometer in diameter and passed at a distance of about two and a half million kilometers.[30] According to scientific calculations a collision between Earth and such an object

> would cause a planetary disaster; at least a *billion* people would be killed, and modern civilization would be destroyed.[31]

In December 1997 an Earth-crossing asteroid with a diameter of almost two kilometers was discovered by astronomers in the United States. Classified as asteroid 1997 XF11, its course was studied closely over the next three months. Then in March 1998 Harvard University astronomer Brian Marsden announced the results of these calculations: There was, he warned, a possibility of a collision in 2028. Headlines on 12 and 13 March of 1998 were dominated by this announcement and astronomers around the world attempted to refine Marsden's calculations. Some concluded that the asteroid would pass closer to Earth than the Moon, perhaps as close as 40,000 kilometers. Others argued that the distance might be more than a million kilometers. Marsden's conclusion was that "the chances of impact are very small, but not impossible." Jack Hills, an asteroid specialist at Los Alamos National Laboratory in the United States commented: "It scares me. It really does. An object this big hitting the Earth has the potential of killing many, many people."[32]

In 1968 the asteroid Icarus, two kilometers in diameter, missed Earth by 6 million kilometers—"an uncomfortably small distance in the scale of the solar system," as the Massachussetts Institute of Technology commented at the time.[33]

In 1991 asteroid BA passed just 170,000 kilometers from Earth, less than half the distance to the Moon. It has a diameter of 9 meters (about the size of a double-decker bus), sufficient "to destroy a small town."[34]

On 16 March 1994, Duncan Steel gave the following briefing to the Australian media:

> About six hours ago Earth had a near-record observed near-miss by an asteroid. The miss distance was about 180,000 kilometers, which is less than half the distance to the Moon. The object is only about 10–20 meters in size. Its name at this stage is 1994 ES1. It was discovered by the Spacewatch team (University of Arizona) at Kitt Peak National Observatory, near Tucson, Arizona. If it had hit Earth it would have done so at a speed of 19 kilometers per second (44,000 miles per hour). Unless it is made of solid nickel-iron (as are many meteorites) it would have exploded in the atmosphere at a height of 5–10 kilometers. The total energy released would be equivalent to a nuclear explosion of energy about 200 kilotons (around 20 times the Hiroshima bomb).[35]

Destructive airbursts caused by asteroids are in fact routinely recorded by the infrared scanners of U.S. military satellites—the recently declassified data for 1975 to 1992 indicates 136 atmospheric explosions with yields of a kiloton or more.[36] One particularly spectacular burst, with a yield estimated at 5 kilotons, was observed over Indonesia in 1978.[37] Even more spectacular was a 500-kiloton airburst between South Africa and Antarctica on 3 August 1963.[38] On 9 April 1984, the captain of a Japanese cargo plane reported a brilliant airburst approximately 650 kilometers east of Tokyo. "The blast formed a mushroom cloud rising from 4,267 to 18,288 meters in only 2 minutes."[39]

FIREBALLS AND COMETS

On 19 February 1913 a small asteroid entered the earth's atmosphere as a fiery apparition over Saskatchewan, traveling east at a speed estimated at around 10 kilometers per second. It was observed at an altitude of 50 kilometers over Winnipeg and Toronto and over several cities of the northeastern United States. It passed over New York and into the Atlantic. Two minutes later it was observed again over Bermuda.[40] Thereafter all trace of it was lost. It probably fell into the ocean.

In 1972 another fireball was observed in the United States, this time rising up steeply to escape from the earth's atmosphere in which it had only temporarily become enchained. The astronomers L. G. Jacchia and John Lewis calculate:

> It approached with a relative speed of 10.1 kilometers per second and was accelerated to 15 kilometers per second by Earth's gravity as it fell toward the top of the atmosphere. Its point of closest contact to Earth was at an altitude of about 58 kilometers over southern Montana. . . . The body had a diameter of 15 to 80 meters and a mass of at least several thousand metric tons and possibly as high as a million metric tons. It passed within 6,430 kilometers of the center of the earth. If it had passed only 6,410 kilometers from the center of the earth it would have exploded or impacted somewhere in the populated strip of land stretching from Provo, Utah, through Salt Lake City, Ogden, Pocatello, and Idaho Falls. The explosive power would have probably been [equivalent to about] 20 kilotons of TNT.[41]

On 1 February 1994 a bolide entered the earth's atmosphere over the Pacific Micronesian islands, crossed the equator traveling in a southeasterly direction and eventually exploded northwest of Fiji, 120 kilometers above the island of Tokelau. It was calculated to have traveled at 72,000 kilometers per hour.[42] The explosion was blindingly bright and may have had a yield equivalent to 11 kilotons of TNT.[43]

Larger and much faster objects have also approached close to Earth. On 27 October 1890 observers at Cape Town, South Africa, witnessed the apparition of an immense comet, with a tail as wide as a full moon, that stretched across half the sky. During the 47 minutes that it was visible (from 7:45 P.M. until 8:32 P.M.) it traversed about 100 degrees of arc. "Supposing this was a typical small comet," observes John Lewis, "traveling at about 40 kilometers per second relative to the earth, then its observed angular rate of two degrees per minute implies that the comet must have passed within 80,000 kilometers of Earth, about a fifth of the distance of the Moon."[44]

Another fast-moving comet, which streaked across the sky at the rate of 7 degrees a minute, was detected in March 1992 by astronomers at the European Southern Observatory.[45] Its nucleus appeared to be about 350 meters in diameter:[46]

> Again taking the most probable flyby speed as 40 kilometers per second, typical for long-period comets, this comet must have flown by at a dis-

tance of about 20,000 kilometers. Remembering that the diameter of the earth is about 13,000 kilometers, this is very close indeed.[47]

MERCURY

The more we learn about the vast arsenal of projectiles flying around in space, the easier it becomes to understand how neighboring Mars—which may once have offered a congenial home to life—could have been reduced to a tortured and barren hell-world. Indeed, what has happened to Mars is actually the *norm* among the inner planets. It is Earth's continued survival as a functioning ecosystem that seems hard to explain.

Mercury, the innermost planet, is brutally pockmarked with craters and, like Mars, appears to have been stripped of huge segments of its crust:

> Something smashed into Mercury with such violence that its outer layers were torn away and, lost to space, fell into the Sun.[48]

Another characteristic that Mercury shares with Mars—and also with Earth—is the phenomenon of massive craters in one hemisphere being matched by reactive disruption at the antipodal point in the opposite hemisphere. As we have seen, the Martian crater Hellas, which has a diameter of almost 2,000 kilometers, has been connected to a bizarre feature known as the Tharsis Bulge, which is nearly antipodal to it. On Earth the 200-kilometer Chixculub crater in Mexico, the epicenter of the K/T Boundary Event, has been connected to the volcanic scabs of the Deccan Traps in India. In the case of Mercury, NASA photographs show a gigantic crater, 1,300 kilometers in diameter, which has been named the Caloris Basin. Exactly on the opposite side of the planet is an extensive area of "chaotic terrain" where there are no impact craters but where the ground appears to have been smashed to bits by gigantic pile-drivers and then shaken up into a new and extraordinary configuration. Duncan Steel offers this explanation:

> When Caloris was formed, huge seismic waves were focused through the interior of Mercury, meeting at the antipodal point and breaking up the smooth terrain that previously existed there.[49]

VENUS

If in our imaginations we look down on the solar system from above, that is, from the north, we will see that all the planets are orbiting the Sun in a counterclockwise direction. The majority of them also rotate counterclockwise about their own axes. The notable exception is Venus, the second planet out from the Sun, which rotates in the direction opposite to its revolution.[50]

Astronomers regard the retrograde rotation of Venus as "quite remarkable."[51] The generally accepted explanation is that at some point in its history it "was struck so hard"—probably by a titanic asteroid or comet—that its rotation was momentarily halted and that it then "began to spin in the opposite direction."[52] The cataclysm is thought to have taken place billions of years ago, during the early stages of the formation of the solar system, but there is also evidence of a much more recent giant impact:

> The entire surface of Venus was wiped clean. . . . Geologists describe this event as having "resurfaced" the planet with lava from its interior as great blocks of the surface cracked and subsided.[53]

EARTH

Earth is the third planet out from the Sun—a glowing sphere of light and consciousness soaring in dark space, a kind of magic, a kind of miracle. Some see it as a living being. Plato described it as a "blessed god . . ."[54]

> a single spherical universe in circular motion, alone but because of its excellence needing no company other than itself, and satisfied to be its own acquaintance and friend.[55]

It is also, with our as yet extremely rudimentary knowledge of our cosmic environment, the only place in which we can be absolutely certain that life exists. The balance of probability is that there is life, perhaps much more intelligent than ourselves, on other planets orbiting other suns. *But we just can't be sure.* For all we know cosmic smashups like those that ruined Mercury, reversed the rotation of Venus, and flayed the planet

Mars may be commonplace not only within the solar system but in the universe as a whole.

Imagine the responsibility, therefore, if we are the only life. Imagine the responsibility if our spark of consciousness is the only consciousness that has survived in the entire universe. Imagine the responsibility if some avoidable threat is looming which through complacency we do nothing about.

JUPITER

What is already clear is that Earth is at present the only planet *in the solar system* that is inhabited by intelligent beings. This may not have been true 10,000 or 20,000 or 50,000 years ago—who knows?—but today all our neighbors are dead and show signs of having suffered massive bombardments of cosmic debris.

Mercury is dead. Venus is dead. The Moon is dead. Mars is dead. And although Earth still lives, with us upon it, there is no evidence that the bombardments have stopped just because we are here. On the contrary, as recently as 1994 humanity was offered spectacular proof that objects of world-killing size do still collide with planets. That was the year in which a swarm of massive fragments from the disintegrating comet Shoemaker-Levy 9 hit Jupiter, an event taken by many astronomers as a timely reminder that Earth, too, could suffer such a fate—and theoretically at any time. As David Levy, the co-discoverer of the comet, observed:

> It was as if Nature had called over the phone and said "I'm going to drop 21 comets on Jupiter at 134,000 miles an hour. . . . All I want you to do is watch.[56]

The impacts were watched—with great interest and attention. Dozens of observatories and the Hubble space telescope, as well as the NASA probe *Galileo,* focused their attention and cameras almost exclusively on Jupiter during the month of July 1994 when the collisions took place, and ominous photographs of all the major impacts were broadcast as headline news to billions of people around the world.

Mercury . . . Venus . . . the Earth-Moon system . . . Mars . . .

Jupiter is the fifth planet out from the Sun; its orbit lies about 500 million kilometers beyond that of Mars. With a diameter of nearly 144,000 kilometers, it is the giant of the solar system—one-tenth of the size of the Sun itself, ten times larger than Earth and 20 times larger than Mars. Its surface is not thought to be solid, but fluid and gaseous, "composed mainly of hydrogen and helium in near-solar proportions."[57] Nevertheless its mass is 318 times greater than that of Earth and, indeed, greater than the combined mass of all the other planets in the solar system.[58]

The ability of such a leviathan to shoulder aside or destroy objects approaching it from space, and to absorb the impacts of those that penetrate its atmosphere, seems virtually limitless. And yet Jupiter was horrifically battered and bruised by its high-speed encounter with the 21 fragments of comet Shoemaker-Levy 9.

COSMIC TRACER

Caroline Shoemaker, the late Eugene Shoemaker, and David Levy discovered their eponymous comet on 24 March 1993. It initially showed up as a fast-moving smudge on grainy photographic plates. Big observatories then turned their telescopes on the object, and Jim Scotti of the University of Arizona's Lunar and Planetary Laboratory, using the 90-centimeter Spacewatch telescope, was the first to confirm that S-L 9 was not in fact one object but "a string of 21 fragments."[59] Early photographs showed images that were beautiful but scary—like tracer bullets arching across the night sky—and astronomers began to calculate how large the individual fragments might be, where they had come from, and where they were going.

It quickly became apparent that the 21 nuclei in the S-L 9 string had all originally been part of a single, much more massive comet, probably between 10 and 20 kilometers in diameter.[60] The largest fragment was estimated at 4.2 kilometers in diameter and others at 3 kilometers and 2 kilometers in diameter.[61] As astronomers plotted their course and calculated their orbit backward it was discovered that "these nuclei had made a very close passage by Jupiter in July 1992."[62]

Further investigations showed what must have happened: the original comet had approached too close to Jupiter, falling to an altitude of just 20,000 kilometers above its surface on 7 July 1992 and breaching the planet's Roche limit. David Levy describes the effects this way:

> Like a giant hand reaching up and pulling the comet apart, Jupiter's gravity pulled on the closest part harder than it pulled on the most distant. As the comet started to stretch out like a noodle, with a shudder it simply became unglued.[63]

Only narrowly managing to avoid collision at that time, it seems that S-L 9 was torn out of its own long-distance orbit through the solar system by this encounter and forced instead into a perilously close-orbit around Jupiter.[64] By mid-May 1993 astronomers had calculated that this orbit would bring the 21 fragments into an even closer encounter sometime in July 1994.[65] Further calculations revealed that this next encounter would be so close that a collision was inevitable:

> Although the comet fell apart in 1992, its pieces survived the graze with Jupiter, but only to buy a little time. The ancient comet would have one orbit left, a last chance to swing away from Jupiter, look back, and return again to crash into the planet.[66]

COMETS REALLY DO HIT PLANETS

Traveling at a speed of 60 kilometers a second, fragment A—one of the smallest—hit Jupiter on 16 July 1994 creating a gigantic plume of fire. A few hours later, fragment B, surmised to be a "loosely held-together group of dust and boulders,"[67] produced a faint plume that lasted for 17 minutes.[68] Two impacts separated by an interval of an hour were associated with fragment C, closely followed by a "short-lived fireball" associated with fragment D.[69] The first large fragment was E. It hit at 11:17 Eastern Daylight Time, sending up a plume of material "more than 30 times the brightness of Europa" (one of Jupiter's moons).[70] As the initial atmospheric turbulence subsided it became clear that the fragment had opened up three huge scars in Jupiter's swirling surface—including one bright spot with a diameter of more than 15,000 kilometers.[71]

Fragment F produced an even bigger impact scar with a diameter of 26,000 kilometers. Then, recounts David Levy, "the gates of hell opened as the central mass of fragment G blew up, leaving a mighty fireball soaring some 3,000 kilometers above the clouds."[72] The fireball rose at 17 kilometers per second and was fueled by superheated gas—twice as hot as the surface of the Sun.[73]

The impact ring created on Jupiter's surface by fragment G was an equally turbulent feature. It expanded outward at the rate of 4 kilometers per second and soon reached a diameter of 33,000 kilometers[74]—just 7,000 kilometers less than the equatorial circumference of Earth. Within another hour it had grown into a spot so big that it could have swallowed Earth, and so bright that it outshone Jupiter's own radiance and temporarily "blinded" telescopes.[75]

"I began to think about what all this meant," remembers Gerrit Verschuur:

> Given that fragment G was supposed to have been 4.2 kilometers across, and given that it was traveling at 60 kilometers per second, its impact energy would have been about 100 million megatons of TNT, something like the K/T impactor that wiped out the dinosaurs. And there it had happened on Jupiter in 1994! What now were the odds on it happening here? The impact produced the equivalent of 5 million Hiroshima-sized explosions going off simultaneously. Incredible! It wasn't so long ago, back in 1991 at the First International Symposium on Near-Earth Asteroids in San Juan Capistrano, California, that I had heard it predicted that we would never see objects of this size slam into planets in our lifetimes.[76]

Gene Shoemaker was asked what he thought was the most important lesson learned from S-L 9. "Comets really do hit planets," he replied.[77]

In an interview with the BBC in London, Caroline Shoemaker was asked to describe what would happen if a fragment like G were ever to hit the earth. Her reply was brief and to the point: "We would die."[78]

20

Apocalypse Now

BY the time all 21 fragments of comet S-L 9 had buried themselves in the massive body of Jupiter, many people who had previously taken little interest in the sky began to look heavenward with feelings of vague anxiety. It took no more than common sense to realize that what had happened to Jupiter could just as easily have happened to Earth—and probably would one day. An old idea of using nuclear missiles to divert potentially dangerous comets or asteroids was revived and there was talk of adapting "Star Wars" technology to defend the earth. It was of course not an accident that only two days after the armageddon-like impact of fragment G, the House of Representatives wrote a clause into the NASA Authorization Bill (quoted in the last chapter) instructing the agency to "identify and catalogue the orbital characteristics of all comets and asteroids that are greater than 1 kilometer in diameter and are in an orbit around the Sun that crosses the orbit of the earth."

SPEED ENERGY

Studies have been done of the possible consequences for the earth, and for human civilization, of collisions with various types and sizes of asteroids and comets. In order to grasp the results of these studies it is important to remember that with impactors of more than a few tens of meters in diameter such collisions will *inevitably* have catastrophic effects—witness, for example, the devastation caused by the Tunguska object in 1908.[1]

The reason is that these projectiles carry huge reservoirs of kinetic energy (the energy of motion of a body or system equal to the product of half its mass and the square of its velocity), which they surrender explosively, generating terrific shock waves, as they snowplow through the atmosphere.[2] Then comes the smash with the planet's surface that deposits sufficient residual energy as heat to melt or vaporize both the impactor and "an amount of target material whose mass ranges from 1 to 10 times the mass of the impactor as the impactor speed increases from 15 to 50 kilometers per second."[3]

Coming in somewhere in the middle of this speed range at 20 to 30 kilometers per second, although velocities as high as 72 kilometers per second have been recorded,[4]

> an asteroid will be brought to a halt in a distance about equal to its own diameter, being literally turned inside-out in the process. Pressures of several million atmospheres and shock temperatures of tens of thousands of degrees are immediately generated.[5]

BIG LAND IMPACTS

Projections have considered the implications of impacts both on land and in the oceans. Professor Trevor Palmer of Nottingham Trent University in England paints this picture of the first effects of a 10-kilometer object striking land at about 30 kilometers per second:

> Bolide and rock would be instantly vaporized, and a crater about 180 kilometers in diameter would be formed within seconds. If, for example, the bolide hit Milton Keynes, the crater would stretch from Nottingham in the north to London in the south, and include Birmingham, Oxford, and Cambridge. This huge crater would be lined with molten rock, and an intense fireball would rise through the atmosphere, producing a violent, scorching wind.[6]

Dr. Emilio Spedicato of the Department of Mathematics and Statistics at the University of Bergamo in Italy reports that the atmospheric disturbance resulting from collision with a 10-kilometer object

> would be colossal and extend over hemispheric areas. For instance, it can be estimated, if ten percent of the initial energy goes into the blast wave, that at 2,000 kilometers from the impact point the wind velocity would be 2,400 kilometers per hour with a duration of 0.4 hours and the air temperature increase 480 degrees. . . . At 10,000 kilometers these numbers would be respectively 100 kilometers per hour, 14 hours, and 30 degrees.[7]

Victor Clube of the Department of Astrophysics and Applied Mathematics at Oxford and Bill Napier of the Royal Armagh Observatory have calculated that if such an impact were to occur in India it would "flatten forests in Europe, setting them ablaze."[8]

> Debris thrown out of the crater would range from mountain-sized lumps, themselves formidable missiles, to hot ash thrown worldwide and adding to the incineration below. Earthquakes would be felt globally and would everywhere be at the top end of intensity scales, with vertical waves many meters high and horizontal ones (e.g., push-and-pull waves) or similar amplitude. These waves would run around the world for some hours.[9]

An immediate effect of the impact would be the simultaneous explosion of "hundreds of fires over an area about the size of France."[10] These would rapidly merge into a single vast conflagration and at least 50 million tons of smoke would be ejected upward, rising to an altitude of 10 kilometers.[11] Within just a few days, fanned by residual windstorms, the wildfires would spread around the globe[12]—as we know actually did happen 65 million years ago in the K/T Event.[13] The pall of smoke would mix promiscuously with the estimated 100,000 cubic kilometers of floating ash and dust thrown into the upper atmosphere by the original impact.[14] With the loss of sunlight, land temperatures would plummet to Siberian winter levels, thick ice would form over rivers and lakes, animal and plant life would be devastated, and all farming would cease.[15]

Another inevitable consequence of any very large land impact would be chemical changes in the atmosphere. According to Professor Palmer: "The fireball would fuse atmospheric nitrogen and oxygen to form nitrogen oxides, which would later react with water to form nitric acid. Similarly sulphuric acid might be produced from burning plant material."[16] Spedicato calculates that such reactions "would completely remove the protecting layer of stratospheric ozone."[17] As the sky gradually cleared of smoke, ash and dust, therefore, any surviving creatures on Earth would be exposed to "ultraviolet radiation of germicidal intensity."[18]

The above calculations assume that the impacting asteroid or comet would enter the atmosphere at a fairly steep angle. But if the angle were shallow, additional complications would ensue. Peter Schultz of Brown University and Don Gault of the Murpheys Center of Planetology have looked into the implications of a 10-kilometer object traveling at 72,000 kilometers per hour striking the earth's surface at an angle of less than 10 degrees from the horizontal. They note that such an impact would be unlikely to produce just one large crater. Instead the bolide

> would break up into a swarm of fragments ranging in size from a tenth of a kilometer to a kilometer in diameter. The fragments would ricochet downrange [and would] eject enough debris into orbit to give the Earth a ring like one of Saturn's.

Over the following two or three thousand years, large chunks of this debris—with estimated volumes of 1,000 cubic kilometers or more—would reenter the atmosphere and crash back to Earth, sparking off local cataclysms of great magnitude.[19] A shower of such objects could produce a tremendous expanding heat storm and perhaps even spark off a second global conflagration. Duncan Steel calculates that

> at reentry speeds varying from a few kilometers per second up to 11 kilometers per second, 1,000 cubic kilometers of rock will release energy equivalent to about a week's worth of solar energy to the whole planet. In many ways one can imagine the situation as being analogous to a huge griller located at 50 to 100 kilometers above the surface, boosting the surface temperature to over 1,000 degrees C. It is only to be expected that under such circumstances the plant life of the continents would be rapidly dessicated and then ignited.[20]

In summary, at whatever angle a 10-kilometer projectile were to hit Earth, the consequences for humanity would be unspeakably dreadful. It is thought likely that 5 billion people would be killed while perhaps a billion would survive, shell-shocked and disoriented, in scattered pockets all around the world.[21]

SMALL BUT DEADLY

It is obvious that asteroids and comets with diameters of less than 10 kilometers must do less damage on impact. Nevertheless, one of the important lessons learned from comet S-L 9's collisions with Jupiter in July 1994 is that even relatively small fragments can deliver very large amounts of kinetic energy—enough to cause massive planet-wide devastation.

On Earth the impact of a 2-kilometer object would be murderous. "As an absolute minimum," warns Duncan Steel, "we might expect 25 percent of the human race to die . . . with a more likely figure being in excess of 50 percent."[22]

Gerrit Verschuur is convinced that it would not even "take a 2-kilometer object to plunge us back into a dark age. . . . It now seems fairly certain that a half-kilometer object would do nicely."[23] Trevor Palmer is of the same view. He points out that an impact with an object 0.5-kilometer wide would release energy "equivalent to about 10,000 megatons of TNT, which is half a million times greater than the energy of the atom bomb dropped on Hiroshima in 1945. For a 1-kilometer asteroid, whatever its composition, the impact energy [which rises disproportionately to size] could be greater than a million megatons"[24]—roughly equivalent in explosive power to the world's entire stockpile of nuclear weapons being detonated all at once.[25]

It is mind-boggling to consider the consequences of a swarm of 10,000-megaton impactors hitting the earth. In built-up industrial areas the fire and blast damage would be enormously complicated by the presence of gas and fuel depots, which would explode like huge bombs. Other flammable chemicals would ignite, releasing plumes of noxious smoke, nuclear power stations would go into meltdown, and ammunition dumps would blow sky-high. Even at great distances from the impact people in downtown areas would be horrifically lacerated—and tens of thousands would be killed—by shards of flying glass (more than 90 percent of all casualties in the London Blitz during World War II were caused by flying glass).

In areas where any large concentration of people survived it is not difficult to imagine how many would be injured, or sick, or poisoned, or burned, or starving, or hypothermic, or insane, or threatened by marauding bands of hungry killers. Nor, when all this is taken into account, is it difficult to realize how quickly and completely the emergency services would be overwhelmed—assuming the emergency workers, their vehicles, and their equipment had themselves survived. It is probably true to say that the fire, police, and ambulance services of most industrialized countries are already overburdened, and that even in "normal times" any concentration of emergencies over a period of days could bring the entire system close to total collapse. A series of 10,000-megaton explosions would produce emergencies on a scale never before seen or imagined and would plunge the world into a nuclear winter.

But if the prognosis is bad for the rich, high-tech industrialized Northern Hemisphere it is perhaps even worse for the low-tech, impoverished, overpopulated Southern Hemisphere. Duncan Steel believes that many Third World countries would simply be wiped out:

> They have neither the advanced agricultural capabilities nor the food stores to survive through a period of duress; witness the famines that occur in Africa during every drought.[26]

IMPOTENCE

The story of famine in Africa in the second half of the twentieth century is a testament to the abject failure of the community of nations to intervene successfully in quite small and local natural disasters that ought to have been swiftly and easily resolved.

Another example to bear in mind is Britain's lengthy indecision and procrastination over the resettlement of the 12,000 inhabitants of Montserrat, the tiny Caribbean island drowning under a relentless tide of lava and ash from its own volcano. Rescues on this scale, and far, far larger, might have to be staged thousands of times over if Earth were ever struck by a series of 10,000-megaton projectiles.

During 1997 much of Southeast Asia fell under a dense cloud of acrid and choking smog—so thick at times that several aircraft crashed, schools and factories had to be shut down, and hospitals registered a huge upsurge in respiratory complaints. The "haze," as it was called, was caused by fires raging in a few thousand square kilometers of Indonesian rainforest. For many months, however, neither the Indonesian government nor neighboring Singapore and Malaysia—nor the world at large—took any effective action to put these fires out and prevent further ones from starting.

Such impotence in the face of extremely damaging environmental and economic threats suggests how little humanity might actually be able to do in the event of a major land impact. Yet in many respects the impact of an asteroid or comet in one of the world's oceans could be far worse.

OCEANIC IMPACTS

In March 1993, Jack Hills and Patrick Goda of the Los Alamos National Laboratory in New Mexico published a research paper in the *Astronomical Journal* arguing that "waves caused by open ocean impacts may be the most serious problem produced by impacting asteroids short of massive killers such as the Cretaceous-Tertiary impactor."[27] In the paper they present disturbing evidence:

> An asteroid with a radius of 200 meters that drops anywhere in the mid-Atlantic will produce deep-water waves that are at least 5 meters high when they reach both the European and North American coasts. When it encounters land, this wave steepens into a tsunami over 200 meters in height that hits the coast with a pulse duration of at least 2 minutes. . . . A disproportionate fraction of human resources are close to coasts.[28]

The wave pulse indicated by Hills and Goda's computer simulations for a 200-meter object would "sweep over all low-lying land, including, for example, Holland, Denmark, Long Island, and Manhattan. Hundreds of millions of people would be wiped out in minutes."[29]

The bigger the impactor the worse the consequences:

> A 500-meter asteroid would produce a deep-water wave 50 to 100 meters in amplitude, even at 1,000 kilometers from ground zero. Since the tsunami height could be amplified by a factor of 20 or more in the

> run-up as continental shelves are encountered, we are referring here to a tsunami several kilometers in height. Even if the impact were between New Zealand and Tahiti, the tsunami breaking on Japan would be perhaps 200 to 300 meters high, and heaven help New Zealand and Tahiti.[30]

Hills and Goda additionally estimate that a 1-kilometer stone object could produce a tsunami *8 kilometers* high. And if the impactor were made of iron it is theoretically possible that the tsunami could reach a height of 28 kilometers.[31] "These numbers," observe the two scientists, "are very disturbing. . . . Perhaps the legendary tale of the lost civilization of Atlantis . . . was due to such a tidal wave."[32]

LONG WAVES BECOME HIGH WAVES

Why is it that oceanic impacts of cosmically rather small objects can produce such enormous waves?

The Japanese word *tsunami* means "harbor wave." These phenomena, normally produced by suboceanic earthquakes, are experienced frequently in Japan and throughout the Pacific region. The great Chilean earthquake of 1960, for example, produced a tsunami that pounded Hilo in Hawaii and parts of the Japanese coast 16,000 kilometers away.[33]

What happens is that the earthquake stirs up waves that are extremely long but very shallow:

> On a ship at sea one would scarcely notice the swell . . . but approaching a shoreline a wave slows down and increases in amplitude as it enters shallow water. There is a piling-up of water as the forward part of the wave slows down.[34]

The experts say that precisely the same effect, *magnified many times over,* would be produced by an impacting asteroid or comet and that the long, seemingly gentle waves that it would produce in the unconstrained environment of a deep ocean would on contact with coastlines rear up into prodigious tsunamis capable of flooding entire continents and destroying everything in their path.

The largest oceanic impacts would have particularly horrific consequences. Crater expert Don Gault has considered the effect of a 10-kilometer object and concluded that in water it would produce a temporary, approximately hemispherical "crater" with a maximum depth of 13 kilometers and a maximum diameter of 30 kilometers.[35] Emilio Spedicato recounts the sequence of events:

> Most of the available energy (92 percent) would be spent in ejection of water, shock heating and formation of waves, the remaining being transformed into potential energy of the displaced water. The formed crater would soon collapse, a column of water 10 kilometers high developing over the impact point. The final collapse of the column originates a system of waves, with amplitudes decreasing, in free ocean, inversely with the distance. The height of the waves would be about one kilometer at 10 kilometers from the impact and one hundred meters at 1,000 kilometers. On approaching the shores substantial amplification of the wave height would follow, the exact value of the amplification depending strongly on the geometry of the coast. In any case, a global catastrophic tsunami, with substantial continental flooding, would be a consequence of an oceanic impact.[36]

Since the average depth of the world's oceans is only 3.7 kilometers[37] it follows that objects 10 kilometers in diameter would hit the ocean bottom with much of their kinetic energy still intact.[38] The implication, if such an object were to fall in an ocean 5 kilometers deep in an area where the ocean crust is also 5 kilometers deep, is that about 35 percent of the transient cavity would be excavated in water, 25 percent in oceanic crust, and 40 percent in the underlying mantle.[39] Researchers Emiliani, Kraus, and Shoemaker agree with Gault and Spedicato that "monstrous gravity waves with heights of several hundred meters" would be produced by such an event and would roll for thousands of kilometers across the world ocean. They, too, believe that the resulting "super-tsunamis" would penetrate deeply into the surrounding continents[40]—as do Victor Clube and Bill Napier, who have presented evidence that a 10-kilometer oceanic impact "would create a hydraulic bore of awesome dimensions and a deep and catastrophic inundation of the land."[41]

WOUNDS

Mercury . . . Venus . . . the Moon . . . Earth . . . Mars . . .

With the exception of Earth, which has survived despite a series of tremendous batterings, we now know that all the other large bodies in the inner solar system—all of them, without exception—have been utterly devastated by cataclysmic impacts of cosmic debris. Among them Mars was once by far the most Earth-like—possessing great oceans and rivers, abundant rainfall, and a dense, quite possibly breathable atmosphere. Yet all this was torn from it in an instant and, it would seem, with the utmost violence. As we saw in part 1, our neighboring planet still bears the wounds of the killer impacts that destroyed it and of the tidal waves, kilometers high, that scoured its surface at the moment of its death.

Scientists for a long while believed that most of the impact craters and other damage visible on Mars must have been inflicted billions of years ago, that the solar system today is a far quieter and far safer place than it was in primordial times, and that the chances of Earth colliding with an asteroid or comet are so small as to be insignificant.

We now know that they were wrong about Earth—and new evidence, which we will review in the next chapter, has forced the abandonment of the formerly dominant uniformitarian view. Could they also have been wrong about Mars? And could there indeed be some kind of mysterious connection between the two planets, as so many ancient sources seem to suggest?

21

Earth Cross

EVERYTHING is moving. Nothing stays still.

The Moon moves around its own axis and orbits Earth. Earth moves around its own axis and orbits the Sun. The Sun moves around its own axis and orbits the center of the galaxy. And the galaxy too is in motion through the expanding universe.

Earth is our home, and our immediate concern. But we will see that it is subject to mysterious and violent tides that perturb the entire solar system and that are governed by the galaxy. If we wish to have a clear picture of what it means to live on this planet, we are obliged to take account of the galaxy and the solar system, and we would be wise to pay attention to any lessons that neighboring planets have to teach. After all, we share their cosmic environment so closely that whatever happens to them can reasonably be expected to happen to us.

Mercury, Venus, the Moon, Mars, and Jupiter all tell us one thing, very simply and very clearly. In Gene Shoemaker's words: "Comets really do hit planets."[1]

And, as we shall see, not only comets hit planets (although comets are by far the most deadly danger), but also vast swarms of meteoroids and asteroids, ranging in size from a meter up to 1,000 kilometers, tear through the solar system at furious speeds.

Such objects, in all possible size ranges, can and frequently do hit planets. Earth has not encountered a very big one—say in the 200-kilometer-plus range—for billions of years. But we now know that it has encountered several in the 10-kilometer range in just the last 500 million years, and that each of these collisions has resulted in the near total extinction of life.

To find out what Earth would look like if it had taken direct hits from a barrage of much bigger objects, we only need to look at the ravaged face of Mars. Curiously, when we do so, we find a "face" staring back at us from the plains of Cydonia.

CROSSING THE LANES

If we envisage the orbits of the planets as a series of flat circular lanes laid out concentrically around the Sun, little Mercury turns in the inner circle. Outside it is Venus, then the Earth, then Mars, then Jupiter. Beyond Jupiter, far from warmth and light, are four farther planets—Saturn, Uranus, Neptune, and Pluto, respectively. Circulating among them all, crisscrossing the lanes in which the planets move, are the turbulent swarms of orbiting rocks and iron we have discussed, loosely classified and graded according to size as either meteoroids or asteroids.

Exactly what these objects are, where they came from, and why some are stony and some metallic (almost like the melted and fused components of gigantic iron machines), are not matters that scientists have settled yet, and there is no consensus. One school of thought is that they are the left-over debris of the iron core and stony mantle of an exploded planet.[2] However, no convincing mechanism has yet been suggested to explain how a planet-sized body could explode. Another idea is that they are remnants from the early days of the solar system—the surplus matter not used up in the formation of the planets. A third theory, the one that we ourselves favor, is that they are closely related to comets, particularly to giant interstellar comets that periodically enter the solar system. The argument is that many of the asteroids and the smaller meteoroids may be the fragmented remains of these dead comets.

BIG UNSTABLE OBJECTS

Fully 95 percent of all known asteroids lie in the "main belt" between the orbits of Mars and Jupiter. But several other populous groups of asteroids circulate between the orbits of Mars and Venus—straddling Earth's orbit.

These are thought to have been "the principal producers of craters larger than 5 kilometers in diameter on Earth, Moon, Venus, and Mars."[3]

There are also large asteroidal objects that lie permanently outside the orbit of Jupiter and others, with highly elliptical orbits, that cross Jupiter's path as they climb toward aphelion (farthest point from the Sun) but that swing into the domain of the inner planets as they fall toward perihelion (closest to the Sun).

Among these asteroidal objects is 944 Hidalgo, which has an orbit of 14 years and a diameter in the range of 200 kilometers. On each turn that it takes around the solar system it swings out far beyond Jupiter—almost as far as Saturn—and then swings back in again approaching the orbit of Mars.[4]

Another more distant and probably slightly bigger object (estimates vary from 200 kilometers to 350 kilometers[5]) is 2060 Chiron, which presently orbits between Saturn and Uranus but has exhibited highly unstable behavior in recent years.[6] Astronomers studying its trajectory have concluded that it is very likely in due course to fall into the inner solar system and perhaps to become an Earth crosser.[7] If that were to happen, says Duncan Steel, it

> would spell disaster for humankind even if Earth did not receive an impact by Chiron itself, or even any large lumps, because the amount of dust in the atmosphere would lead to a significant cooling of our environment.[8]

A third 200-kilometer-plus asteroid is 5145 Pholus.[9] Its steeply elliptical orbit takes it across the paths of Saturn, Uranus, and Neptune.[10] Like Chiron, it has been described by astronomers as "inherently unstable" and is thought likely to "plunge into an Earth-crossing orbit"—although probably not soon.[11]

There is a frightening object called 5335 Damocles, estimated to be 30 kilometers in diameter, which crosses the orbit of Mars at perihelion and then swings out as far as Uranus before returning to the inner solar system again in an orbit of forty-two years. According to Duncan Steel of Spaceguard Australia:

> This asteroid has an elongated, high-inclination orbit which would classify it as an intermediate-period comet, except that it shows no signs of

outgassing, seeming to be totally inert. Its name was chosen to remind us of the Sword of Damocles, because its future orbit has a good chance of evolving into an Earth-crossing one.[12]

MAIN BELT

Since the discovery of Hidalgo, Chiron, Pholus, and Damocles, other large unstable asteroids have been found with the same ability to cross from the outer solar system into the inner solar system—and even to threaten Earth.[13] But there are also vast armies of asteroids that revolve around the Sun in stable orbits and present no threat to us at all.

These include the members of the Trojan group that share the orbit of Jupiter, some following the planet, some leading it. Photographic surveys have so far identified 900 individual objects with diameters exceeding 15 kilometers.[14]

All the "main-belt" asteroids orbiting between Jupiter and Mars also appear, for the moment, to be in secure orbits. Their total number is thought to exceed half a million, including such true giants as Ceres.[15] Really a mini-planet in its own right, this country-sized sphere of rock has a diameter of 940 kilometers, revolves around its own axis in 9 hours 5 minutes, and orbits the Sun once every 4.61 years.[16]

Ceres is very dark and reflects only about 10 percent of the sunlight falling on it.[17] To date it is the largest asteroid identified. Next down in size are Pallas (535 kilometers), Vesta (500 kilometers), and Hygeia (430 kilometers). Davida and Interamina are both around 400 kilometers in diameter. Juno is about 250 kilometers in diameter. All in all more than 30 main-belt asteroids with diameters greater than 200 kilometers have now been positively identified and catalogued—with significant new discoveries being made every year.[18]

AMORS

Moving in from the main belt we begin to encounter the first swarms of "near-Earth asteroids," a broad category that includes all asteroids capable of passing inside the orbit of Mars.[19] The most distant of these do not extend as far as the orbit of Earth. But a little closer in there is another family of Mars crossers, the Amors, of much more immediate interest. A characteristic of the Amors (more than 130 had been catalogued by March 1995[20]) is that they are easily perturbed by Jupiter and by our own planet's powerful gravity, with the result that several of them have now changed their orbits to become "part-time Earth crossers."[21] Many others in the same family do not presently approach Earth but, in theory, may be "unpredictably redirected" at any time.[22]

Astronomers from the Observatoire de la Côte d'Azur in France and mathematicians from the University of Pisa in Italy have for some years been paying particular attention to an Amor called 233 Eros, which is 22 kilometers long and 7 kilometers wide—dimensions that make it a substantially bigger and more lethal projectile than the K/T object that killed off the dinosaurs.[23] Although Eros does not currently cross Earth's orbit it does undergo "relatively frequent close encounters with Mars and long-range perturbations by the outer planets."[24] These have altered its course to such an extent that in 1931 it "swished to within 17 million miles of Earth—much closer than any planet."[25] Computer simulations indicate that Eros is very likely to become a true Earth crosser within the next million years and that in the longer term "a collision is likely."[26]

So far about 15 other Amors on Eros-like trajectories have been found, and all of them could one day hit the Earth.[27] None are as massive as Eros, but both 1627 Ivar and 1580 Betulia have diameters approaching 9 kilometers.[28]

APOLLOS

Moving in again from the zone of the Amors we come to the Apollo asteroids (named after 1862 Apollo, a 1-kilometer object, the first in this class, discovered in 1932 by the German astronomer Karl Willhelm Reinmuth).[29] The chief characteristic of the Apollos is that they "deeply cross the Earth's orbit on an almost continuous basis."[30]

Since the early 1990s a number of observatories have mounted aggressive searches to establish the true extent of the "Apollo problem." The conclusions that they have come to are that these Earth-crossing projectiles are extremely numerous, that there are likely to be more than 1,000 of them with diameters exceeding one kilometer,[31] and that some may exceed 50 kilometers in diameter.[32]

Known large Apollos (of which more than 170 had been catalogued by March 1995) include the frightful world killer 2212 Hephaistos, which has a diameter of 10 kilometers.[33] Although smaller, another deep Earth crosser, Toutatis, looks almost equally unpleasant. It is what is known as a contact binary—"two fragments either welded together or held in place by a very feeble gravity."[34] The larger element has a diameter of 4.5 kilometers, while the smaller element is 2.5 kilometers wide.[35] The composite object behaves in an unbalanced and unpredictable manner as it tumbles through space.[36] All that is certain is that it has already crossed Earth's orbital path at a distance from us of just over 3 million kilometers[37]—a distance that our planet covers in about 30 hours—and that the effects of a collision with such a rapidly rotating and unstable object would be devastating.

> The existence of Toutatis proves that there are still giant rocks out there that can be doomsday asteroids and that they come close to us.[38]

Several Apollos with diameters in the 5-kilometer range have been found during the 1990s,[39] and, as we saw in chapter 19, a number of smaller Apollos, such as Asclepius (0.5 kilometers), Hermes (approximately 2 kilometers), and Icarus (2 kilometers), have made extremely close fly-bys of Earth. There are also large and mysterious Apollo objects such as Oljato and Phaeton that behave much more like comets than asteroids, and which we will have reason to investigate in later chapters.[40]

A tiny fragment of Phaeton hit Earth on 13 December 1997. It landed in politically troubled Northern Ireland, close to its border with the Irish Republic, creating an explosion that was initially thought to be a terrorist bomb. Examination of the crater by scientists from the Royal Armagh Observatory and from Belfast's Queen's University showed that it was in fact a meteorite and that the parent body was Phaeton.[41]

It is worth repeating that all of the Apollos are permanently locked in Earth-crossing orbits and that they are accompanied by an unknown number—probably thousands—of as yet undetected and perhaps massive companions. There are no traffic lights at the intersections where they cross the great circle in the sky around which Earth orbits and, over very long periods of time, the laws of chance make collisions inevitable.[42]

Is a collision between Earth and an Apollo object likely at any time in the near future?

The only honest answer to this question is *nobody knows*—because nobody has the faintest idea how many of these projectiles there really are out there. Apollos are notoriously invisible to telescopes and are indeed so elusive that even those that have been catalogued frequently "disappear." The 1862 Apollo, for example, after which the whole swarm is named, was lost to telescopes soon after it was discovered in 1932 and was not spotted again until 1973.[43] Hermes, which passed so close to Earth in 1937,[44] vanished and has not been seen since. For this reason, says Brian Marsden of the Harvard-Smithsonian Center for Astrophysics, it is "one of the most dangerous near-Earth objects."[45] Hephaistos, the biggest Apollo of all, successfully managed to evade detection—despite its 10-kilometer girth—until 1978.[46]

ARJUNAS, ATENS, AND OTHERS

Tom Gehrels, professor of planetary sciences at the University of Arizona at Tucson, and the principal investigator of the Spacewatch program at Kitt Peak Observatory, has identified a special subgroup of Earth-crossing Apollos that he has named the Arjunas. With diameters of up to 100 meters, they follow the orbit of Earth very closely. This means that they are unusually susceptible to our planet's gravitational attraction and have "very short expected orbital lifetimes before colliding with Earth."[47]

Moving in from the Arjunas, the next significant belt of asteroids that we encounter have been named the Atens. Astronomers estimate—although once again it is really just a guess—that at least one hundred of them exceed one kilometer in diameter. They have highly elliptical orbits that put many of them on repeated Earth-crossing paths.[48]

Still further in toward the Sun are other objects following even more steeply elliptical orbits. A typical example is 1995 CR, discovered by Robert Jedicke of Spacewatch in 1995. This 200-meter inner-solar-system wanderer follows

> a highly eccentric path that crosses the orbits of Mercury, Venus, Earth, and Mars. This type of orbit is highly unstable (chaotic) and before long, at an unpredictable time in the future, 1995 CR will smash into one of these four planets, or the Sun, or will be thrown out of the solar system.[49]

Just as scientists cannot give us accurate estimates of when particular asteroids will collide with Earth, or of the absolute numbers of asteroids in any of the subfamilies, so also there can be no firm and final estimate of the total number of potential impactors. A broad consensus has nevertheless been reached by astronomers that there are likely to be at least 2,000 asteroids of a kilometer or more in diameter distributed among the main Earth-crossing families[50] together with somewhere between 5,000 and 10,000 objects of half-kilometer size and perhaps as many as 200,000 objects of quarter-kilometer size.[51] Confirmation of these estimates can only come from close observations of the sky and, indeed, the rate of discovery of Earth-crossing asteroids showed dramatic increases during the 1990s. In 1989 only 49 such objects had been discovered (4 Atens, 30 Apollos, and 15 Amors), but by 1992 this number had increased to 159, an increment of 110 in just two years. Three years later, in 1995, the grand total had risen to over 350, a further increment of 200—making an average for 1989 to 1995 of more than 50 new discoveries a year.

"Although many of these are small objects," commented Duncan Steel in 1995,

> it is true that we have now found many more of the 1-kilometer-plus asteroids that threaten a global catastrophe than we had catalogued only five years ago. However, we still know of only a small fragment of the total population of such objects: few scientists involved in this area believe that we have to date discovered more than 5 percent of that total. Although none of the *known* asteroids is going to hit Earth in the foreseeable future (the next century or two) this is not a particularly comforting fact, because if there *were* an asteroid due to strike home soon, then there is a greater than 95 percent chance that we would not have found it yet.[52]

TIME TO SAVE THE WORLD?

Humanity's fundamental ignorance about the true extent of the threat posed by Earth-crossing asteroids is unlikely to be lifted soon—despite the fact that many scientists seriously believe it would be possible to use controlled nuclear explosions and other techniques to deflect potential impactors if they could be identified in time. It is not our purpose here to explore the various strategies that have been proposed to achieve this objective. Nor would we be in any position to assess their relative merits. Our impression is that many of them are very close to the limits of modern technology. Nevertheless, we have no doubt that the prospect of an imminent collision with a 10-kilometer Apollo would focus the minds of politicians and galvanize global industry and science into action.

But would there be *time* to save the world?

Would there be time to blow up or divert the incoming object, or would it be discovered too late?

Duncan Steel argues that at the present minuscule rate of public expenditure, "it would take perhaps 500 years to complete the search for all the Apollos larger than one kilometer, and longer for the Atens. Thus if one has 'our number' on it for the year 2025, we would most likely not find it ahead of time."[53]

In an official document dated 19 February 1997, NASA notes that "cosmic impacts are the only known natural disaster that could be avoided entirely by the appropriate application of space technology." In the same document NASA then goes on to admit:

> The only technology we have today for defense against asteroids and comets is nuclear, and we would require years of warning in order to deflect or disrupt a threatening object. . . . The truth is that if we found an asteroid headed our way with less than several years' warning, there is nothing we could do to protect ourselves except evacuate population from the impact site.[54]

What would it cost to get those "several years' warning"?

According to a 1991–1992 NASA study, "All potential Earth-impactors down to one kilometer in size could be discovered and tracked in a program costing $300 million spread over five years."[55] A follow-up study, chaired by the late Eugene Shoemaker of Lowell Observatory and completed in 1995, concluded that advances in astronomical imaging systems could allow such a survey to be completed in ten years at a total cost of less than $50 million.[56]

The reader will recall that in 1994 Congress instructed NASA to identify and catalogue all Earth-crossing asteroids greater than one kilometer in diameter within ten years.[57] We were baffled to discover that no such program had been launched by the beginning of 1998 and that NASA's support for asteroid and comet search programs was at that point still limited to about $1 million a year.[58]

The "asteroid threat" remains an underresearched and largely unknown quantity. Assessments of it tend to be complacent—hence, we suppose, NASA's lethargy—and yet such assessments are inevitably founded on the extremely narrow database of present knowledge about asteroids.

How can scientists and governments be sure that the little they have managed to learn so far is not hopelessly unrepresentative of the overall picture?

What level of real certainty is there that Earth is not about to share the dreadful fate of Mars?

In the next chapter we will consider comets, which the Chinese knew as "vile stars."[59] "Every time they appear," wrote Li Ch'un Feng in the seventh century, "something happens to wipe out the old and establish the new."[60]

22

Fishes in the Sea

JOHANNES Kepler, the seventeenth-century astronomer and mathematician, once exclaimed with perceptive wonder, "There are more comets in the sky than there are fishes in the sea!"[1]

We do not know how many fishes there are in the sea, but since the 1950s increasingly refined observations have led astronomers to a mind-boggling conclusion: there are *at least* 100 thousand million (100 billion) comets in the solar system at any one time, stored in two huge reservoirs that are known—after their discoverers—as the Oort cloud and the Kuiper belt.[2]

The Oort cloud, the more distant of the two, lies at the extreme limit of the Sun's gravitational domain, a full light year out—50,000 times the distance between the Sun and Earth.[3] Its form is that of a spherical "shell" entirely enveloping and surrounding the rest of the solar system. A number of astronomers are of the opinion that it may, on its own, contain 100 billion comet nuclei: "Most [are] between 1 and 10 kilometers in diameter, although some may be much larger."[4]

Exactly how much larger, or how plentiful such objects really are, nobody is yet in a position to say; they are too far away from us to be seen by even the most powerful telescopes. It is entirely possible, however, that huge numbers of Oort cloud bodies could be more than 300 kilometers in diameter.

This has already been proven observationally to be the case with comets in the Kuiper belt—a flattened disk-shaped formation that lies beyond the orbit of Neptune. The Kuiper belt is very remote—its outer edge is almost fifty times farther than the distance from the Sun to Earth—yet it is still a thousand times closer to us than the Oort cloud.

Since the 1970s the astronomers Victor Clube and Bill Napier have been developing and refining a theory concerning the occasional penetration and destructive *fragmentation* within the inner solar system of what they call "giant comets," which are hundreds of kilometers in diameter rather than a few tens of kilometers or less, such as those we usually see.[5] While this theory was based on pure logic and calculation, it did not initially receive wide support from other astronomers. Today it is universally accepted. This is because Clube and Napier have been vindicated by telescope observations of the Kuiper belt, which has been proved to contain objects of exactly the sort they had predicted.

The first Kuiper belt object to be detected—1992 QB1—has a diameter of 250 kilometers.[6] Other massive finds include 1993 FW, again about 250 kilometers,[7] and 1994 VK8 and 1995 DC2, which both have diameters of about 360 kilometers.[8] Recent observations have confirmed the impression that such objects may exist in very large numbers. By March 1996 more than thirty of them had been found,[9] and in January 1998 Victor Clube told us that the Kuiper belt is literally "full of giant comets! They're the only things we can see, actually—it's so far away. They're all a few hundred kilometers across."[10] Such discoveries have led to a widely accepted estimate that

> there may be at least 35,000 objects larger than 100 kilometers in diameter orbiting in this region of the solar system just beyond the orbit of Neptune.[11]

Indeed it is a sign of how influential Clube and Napier's work has become that a number of astronomers now consider Pluto, with its unusual elliptical orbit, to be nothing more than an extremely large Kuiper belt object—a former comet that has become a planet. Clyde Tombaugh, who discovered Pluto in 1930, is one of the supporters of this view and now calls it the "king of the Kuiper belt."[12]

COMET-ASTEROID CROSSOVER

Another interesting possibility, which Victor Clube and others have investigated, is that certain large "asteroids" may also be Kuiper belt comets—perhaps in a temporarily "dormant" state—that are gradually falling into the inner solar system.[13] "After about ten million years," explains David Brez Carlisle, "the trajectory of anything orbiting in the Kuiper belt decays into chaos, generally into a quasi-elliptical orbit that [will ultimately bring] it into the zone of the stony planets."[14]

Can comets be asteroids? Can asteroids be comets?

Like so many categories used by scientists, it turns out that the distinction between the two is not clear-cut. From various authorities the notion has entered popular culture that asteroids are formidable rocky obstacles whereas comets are "dirty snowballs." The renowned British astronomer Sir Fred Hoyle strongly disagrees with the second part of this idea:

> Comets are not just dirty snowballs. No dirty snowball at a temperature of minus 200 degrees centigrade ever exploded as comet Halley did in March 1991. Dirty snowballs are not blacker than jet black. On March 30–31, 1986, comet Halley ejected a million tons of fine particles, which on being warmed by the Sun emitted radiation characterized by organic materials, not dirt as one understands dirt.[15]

Whether it is a dirty snowball—or something more—an object is likely to be classified as a comet if astronomers observe that it has the following characteristics:

1. An extremely eccentric (as opposed to more or less circular) orbit, bringing it close to the Sun and then taking it far away again.
2. A volatile chemical composition that produces jets of gas, a large luminous cloud—"coma"—around the frozen central nucleus, and frequently a "tail" consisting of glowing particles blown away from the comet by the solar wind (with the result that the tail always points away from the Sun irrespective of the direction that the comet is traveling in).[16]

With regard to the first characteristic—eccentricity of orbit—new discoveries have revealed a growing number of glaring exceptions to the rule. These include objects that are unmistakably comets in terms of their general appearance and volatility but that nevertheless move in near-circular orbits as asteroids do (the six comets of the Hilda group, for example).[17] Conversely, we saw in chapter 21 that many asteroids have extremely eccentric orbits and that some, such as as Damocles, Oljato, and Phaeton, are already suspected as comets in disguise.

Damocles has "an elongated, high-inclination orbit that would classify it as an intermediate-period comet except that it shows no signs of outgassing, seeming to be totally inert."[18] Phaeton's orbit also has curiously comet-like properties, and during the 1990s the previously dormant Oljato was observed to have become volatile—showing signs of "weak outgassing" and even a faint tail.[19]

Another likely case of mistaken identity among these Earth crossers and near Earth crossers is the 10-kilometer Apollo asteroid Hephaistos, now regarded by increasing numbers of astronomers as a "spent" fragment of a giant comet.[20] Indeed, Victor Clube and Bill Napier maintain that many Apollo asteroids—perhaps most of them—are nothing more than the nuclei of degassed comets or fragments of degassed comets. A typical example is 1979 VA, which "has an orbit like a short-period comet with an aphelion close to Jupiter."[21]

Looking outward to more distant reaches of the solar system, recent observations have demonstrated that the trans-Jovian "asteroid" Hidalgo also has a comet-like orbit.[22] We saw in the last chapter that the trans-Uranian object Chiron has an orbit that is equally hard to label. Observations since the mid-1990s have shown that it is "slightly outgassing" and has begun to release volatiles in a manner that astronomers know is unlike any asteroid.[23]

> Its icy nucleus of 350 kilometers would seem to suggest that it is a giant comet provisionally parked in a quasi-circular but unstable orbit."[24]

For these reasons, says Professor Trevor Palmer, the view that some asteroids may be the remains of former comets is becoming widely held. "This could be the result of an icy nucleus being sealed off completely by the formation of an insulating crust, or by all the volatile material being boiled off, leaving behind a rocky core."[25]

HALLEY'S COMET

The suggestion that 200-kilometers-plus objects like Chiron and Hidalgo could be former comets from the Kuiper belt gradually spiraling down into the inner solar system is supported by observations of smaller comets that have penetrated more deeply. For example, astronomers already agree that the present orbits of periodic comets Halley and Swift-Tuttle must have originated in just such a "spiraling down" after they had been "parked for a few million years in the Kuiper belt."[26] At the extremes of their steeply elliptical trajectories, before plunging back again toward the Sun, both objects still signal their origins by returning to the belt.[27]

"Periodic" comets—the term is a broad one that refers to all comets on orbits that will sooner or later bring them back through Earth's skies—are subdivided by astronomers into three main groups: short-period, intermediate-period, and long-period. Short and intermediate-period comets have orbits varying from less than six years up to two hundred years; long-period comets have orbits of more than two hundred years rising, in some cases, to thousands and even hundreds of thousands of years.[28]

With an intermediate-period orbit of 76 years, Halley's comet last passed by Earth in 1986 at which time it was intensively studied by space probes from several countries. It is a formidable body with an estimated mass of around 80 billion tons and dimensions of about 16 × 10 × 9 kilometers.[29] Its potato-shaped nucleus is extremely black, reflecting only 4 percent of incidental sunlight, and slowly rotates around its axis once every 7.1 days.[30]

Recorded observations of Halley's comet go back more than 2,200 years.[31] Outgassing explosively on each approach to the Sun, it therefore has had the time to scatter immense swathes of debris in its ancient and well-trodden wake. Earth passes through this debris twice each year—in May and in the third week of October—at which times its skies light up with the Eta Aquarid and Orionid meteorite showers that descend from the comet.[32]

THE SWIFT-TUTTLE COLLISION HAZARD

Historical sources and modern observations record the existence of about 450 Earth-crossing comets. Most of these were of the long-period variety and have not yet returned either to menace us or to miss us. Out of the known short- and intermediate-period comets that revisit us more regularly, about 30 are locked on Earth-crossing orbits and could theoretically collide with our planet at some time in the future.[33] Halley's is one of these. Another is Swift-Tuttle, the parent body of the Perseid meteorite shower through which Earth passes each July and August.[34] Astronomers studying Swift-Tuttle's trajectory believe that this comet represents a *serious and imminent hazard.* As it approaches perihelion, its closest point to the sun, computer simulations show that its intersections with the path of Earth can, under certain circumstances, bring it perilously close to us. In particular, and well understood:

> Near collision with Earth would take place if the comet were at perihelion in late July.[35]

For this reason Swift-Tuttle has been described by one authority as "the single most dangerous object known to humanity."[36] Calculations show that it will remain a threat for at least another 10,000 to 20,000 years,

> after which its orbit is likely to deteriorate so that it will either fall into the Sun or be thrown out of the solar system, provided it doesn't hit Earth before it does that.[37]

CAPE EFFECT

The Swift-Tuttle story begins with the first sighting of the comet in July 1862. Over the course of the next month, as it approached to within 50 million miles of Earth, it became a dazzling specter in the night sky with a tail 30 degrees long that was reportedly brighter than the brightest stars.[38] For several weeks it pursued a serene and predictable course through the heavens—a course that was painstakingly tracked and logged by astronomers around the world. During the last few days that it was visible, however, it did something that no comet had hitherto been seen to do: *It changed direction.* As it disappeared from view, the Cape Observatory in South Africa noted with puzzlement that its trajectory had shifted by about 10 arc seconds during its transit of Earth's skies.[39]

This so-called Cape effect is believed to have been caused by outgassing from the comet itself, outgassing so violent that Swift-Tuttle was literally jetted sideways.[40]

But was it a one-shot event, or something that happens regularly? In 1862, questions like these introduced an element of uncertainty into calculations of the likely date of Swift-Tuttle's return—although it was generally felt that the period should be about 120 years.[41] A similar projection was made in 1973 by Brian Marsden, the International Astronomical Union's (IAU) leading expert in the computation of orbits. After carefully rechecking and recalculating the 1862 data he concluded that the comet would return somewhere between 1979 and 1983.[42]

When it did not return on schedule Marsden widened the net of his calculations to include historical observations of comets that could be identified with Swift-Tuttle. He found a close match with sightings from 69 B.C., A.D. 188, and A.D. 1737, and on the basis of these came up with a new estimate that the comet would return in 1992 and would reach perihelion around 25 November of that year.[43]

Marsden's prediction proved to be quite accurate, and the reappearance of Swift-Tuttle—on a trajectory that brought it to perihelion on 11 December 1992—was first observed by the Japanese astronomer Tsusuhiko Kiuchi on 26 September 1992.[44]

THE WARNING

Marsden now returned to his computers with refined orbital information in order to work out the date of Swift-Tuttle's next approach to perihelion. He found that this would occur after a period of about 134 years, on 11 July 2126.[45] Inevitably he began to wonder whether some recurrence of the Cape effect, or other orbital vagary, might cause him to be in error again.

The reader will recall that a near collision between Earth and Swift-Tuttle is to be expected if the comet should ever reach perihelion in "late July"—indeed, it was Marsden who had been responsible for the original calculation that led to that prediction as far back as 1973.[46] Looking at the problem again in 1992, his next step was to work out the exact date in late July 2126 on which a perihelion passage by Swift-Tuttle would be followed by collision with Earth. The computers highlighted 26 July 2126 and indicated that if the comet were to reach perihelion on that day, then it would crash into our planet a little less than 3 weeks later on 14 August 2126.[47]

So, the future of the human race seemed to hinge on the cosmically very small matter of the distance Earth would travel around its orbit in the 15 days between Marsden's calculated perihelion date for Swift-Tuttle of 11 July and the "black-spot" date of 26 July. He had to admit there was a chance he could have missed some vital factor. He therefore issued IAU circular 5636 (October 1992) in which he warned of the possibility that

> periodic comet Swift-Tuttle may hit Earth on its next return.[48]

SAFE FOR THE NEXT MILLENNIUM?

A storm of publicity erupted after Marsden's announcement, and he was accused of sensationalism. Obliged to defend his position, he explained that the purpose of the circular had not been to scare anybody but to urge professional astronomers to pay special attention to the comet "during the next several years":

> The observations in 1862 showed that Swift-Tuttle behaved in a very peculiar fashion—something of the kind I have never seen before in nearly forty years of computing orbits. . . . The fact is that even if Swift-Tuttle doesn't get us next time, it will have ample opportunity to do so in the more distant future.[49]

Marsden spent three months going through all his calculations again. Then at the end of 1992 he made a further statement in which he affirmed that he was now certain that his original date of 11 July would be proved correct—give or take a day or two—and that there was therefore no danger of a collision in 2126.[50] "We're safe for the next millennium," he proclaimed, adding that the comet would make another close approach in the year 3044.[51]

UNCERTAINTIES

Astronomers watching Swift-Tuttle leave the inner solar system observed a recurrence of the Cape effect during 1993: "The comet ejected material that changed its path once again, albeit very slightly."[52] It then continued on its way, traveling so fast that by 1998 the most powerful telescopes on Earth were no longer able to pick it up. It will be seen next when it returns toward perihelion in 2126, with hope closer to 11 July than 26 July.

With a diameter of 24 kilometers, Swift-Tuttle will then be traveling at just over 60 kilometers per second. If by some bad fortune Marsden turns out to be wrong and it does hit Earth, speed/mass calculations indicate that the impact energy will be "in the range of 3 to 6 billion megatons."[53] This would be equivalent to between 30 and 60 impacts on the scale of the K/T event 65 million years ago.

Could there be a collision, or is Brian Marsden's 15-day margin sufficiently wide to save the planet?

It's anybody's guess. As Dr. Clark Chapman of the U.S. Planetary Science Institute observes:

> Astronomers have no idea at this time as to how much the comet's orbit will be shifted due to the disruptive forces working on the comet's surface, which increase as it nears the Sun.[54]

Such uncertainties are characteristic of the entire field of cometary research, where big surprises and big objects constantly materialize out of the darkness of deep space. Although the odds are imponderable, it should be obvious even to a schoolchild that Swift-Tuttle could go on missing Earth forever, and that another comet, perhaps one that has not been seen in our skies for thousands of years, could materialize tomorrow threatening our doom like the dragon of Revelations,

> which had seven heads and ten horns. . . . Its tail dragged a third of the stars from the sky and dropped them to the earth.[55]

Little wonder then, when the very bright, long-tailed, long-period comet Hale-Bopp appeared ominously in 1997—making its closest approach to Earth at the spring equinox after not being seen for an estimated 4,210 years—that a sort of eschatological fever briefly seized the world. Moreover, if Hale-Bopp had hit us instead of passing us by at a distance of 200 million kilometers it really would have been the last of our days. This comet is thought to be at least twice the size of Swift-Tuttle.[56]

SNEAKING UP

Other long-period comets with orbits of 15,000 years, or 20,000 years, or 90,000 years, could theoretically appear out of the night sky *at any time—without any warning*. Since their previous visits are recorded in no known historical documents or traditions, we have no way of predicting when they will be coming back. The same goes for long-period comets that may have passed this way in historic or near-historic times—like Hale-Bopp in 2210 B.C.—but for which, again, no record has survived.

Such comets, say Philip Dauber and Richard Muller, are "as likely to be orbiting the Sun opposite to Earth's direction as with it." When this happens,

> their potential impact speeds are even greater than those of short-period projectiles. Their usually large size—4 kilometers and up—makes them still more hazardous. These Earth-crossing comets only become visible as heat from the Sun begins vaporizing their long-frozen ices. . . . About a year of acceleration remains before they swing around the Sun or, rarely, collide with a planet. About half of all long-period comets are actually Earth crossers. . . . If we are especially unlucky, a new comet on a collision course with Earth could be detected with only two months remaining before the fatal crash.[57]

David Morrison of NASA's Ames Research Center points out that with present technology "no means exists to distinguish a faint object (either comet or asteroid) against the dense stellar background in the Milky Way."[58] He warns that it is therefore

> possible for a comet to "sneak up" on Earth, escaping detection until it is only a few weeks from impact. A perpetual survey is required to detect long-period comets, and even with such a survey we cannot be sure of success.[59]

WHAT SCIENCE REALLY KNOWS

It seems that a process of evolution is at work in the life of comets and that long-period comets gradually change their orbits through "the buildup of gravitational interactions with the major planets"[60] to become intermediate-period comets and finally short-period comets with shorter and shorter orbits. So short, eventually, that they must either fall into the Sun or become enchained in the gravity of a planet. An example is Encke's comet, an Earth crosser, which has the shortest period of all known comets—just three and one-third years—and which has been observed to become "increasingly erratic in keeping its appointments in our skies."[61] The period of its orbit is shortening fast and, as we will discover, it may be part of a larger conglomeration of cosmic debris that is presently evolving into a deadly collision hazard.[62]

In the past two centuries two particularly near misses have been recorded between Earth and comets. Comet Lexell missed Earth by less than a day in June 1770,[63] and comet IRAS-Araki-Alcock flew by at a distance of about 5 million kilometers in 1983.[64]

When can the next close approach be expected?

The classic work of reference on comets, to which all scientists seeking guidance on these matters automatically turn, is Brian Marsden's *Catalogue of Cometary Orbits*. The 1997 edition lists all of the 1,548 comets for which sufficient data exist to compute orbits—91 from the extremely scanty historical data that has come down to us from the period before the seventeenth century and the rest "from cometary passages during the last three centuries."[65]

What science really knows about comets, in other words, derives from data based on an incredibly narrow sample of cometary behavior as observed from our tiny corner of the universe in three insignificant centuries.

FRAGMENTING GIANT COMETS

We have seen that countless billions of comets are in the Oort cloud and the Kuiper belt, that some of these comets seem to be "spiraling down" toward the Sun—and thus toward the inner planets—and that many objects previously believed to be asteroids are in fact the remains of former comets. In a sense, therefore, it is no longer useful to think of asteroids and comets as distinctly different objects. Instead they look like the consequences of an hierarchical disintegration process in which giant comets from the outer solar system with very long orbits migrate into the inner solar system fragmenting along the way into a multitude of smaller shorter-period comets, which in turn either collide with planets—chemical tests indicate that the K/T impactor was an active comet—or manage to avoid doing so.[66] Those that survive will put on ever-diminishing firework displays of dust, meteorites, and larger debris for a few thousand years before eventually becoming completely devolatilized and inert—that is, comets in asteroidal form. They do not lose their propensity to fragment, however, nor to bump into planets, and continue to cross orbits with the random danger of a game of Russian roulette.

As we have seen, it is only since the mid-1990s that the fragmenting "giant comet" idea, which was vigorously advocated by Victor Clube and Bill Napier more than twenty years earlier, has begun to win universal favor among astronomers. The discovery of huge comets like Chiron and Hidalgo, as well as the Kuiper belt objects, has settled that. Moreover it is now clear from a study of historical records that giant comets do not always fragment in the outer solar system and can sometimes survive, more of less intact, to approach the domain of the inner planets. A notable example was comet Sarabat in 1729 that almost reached Jupiter.[67] From a number of astronomical reports made at the time it is known that this comet was extremely bright—"intrinsically the brightest observed in recent centuries," says Duncan Steel,[68] that "only a very large object could have appeared so bright when so far away,"[69] and that

> a lower estimate of its size is about 100 kilometers; actually it might have been up to 300 kilometers across. . . . It is inevitable that many similar comets on Earth-crossing orbits have arrived over geological time.[70]

To this Bill Napier adds that 200-kilometer objects in chaotic orbits are inherently unstable: "It only takes a small collision to veer a comet on a path toward Earth, and who knows what it could do?"[71] Such unpredictability is of course heightened by the distinct possibility that many comets may also be subject to Cape effects because of outgassing. In the case of Halley's comet an accurate estimate of the power of these gas jets was obtained by the *Giotto* space probe. The jets were found to

> exert a force of about 5 million pounds, or nearly as much as all the engines of the space shuttle as it lifts off from the launch pad. And these jets continue for hour after hour, day after day.[72]

MULTIPLE INDEPENDENTLY TARGETED REENTRY VEHICLES

Since the first optical confirmation of the existence of giant comets in the Kuiper belt in 1992 no such object has yet been seen to fragment. "Ordinary" comets, however, which are intimately related to the giants in every respect, are frequently observed to break apart releasing swarms of "warheads"—like MIRVed intercontinental ballistic missiles.

One example was comet Biela, which had a computed orbit that came "within 20,000 miles of the Earth's."[73] (Although this of course does not mean that Earth and the comet were ever actually within 20,000 miles of each other; that would depend on where each of them were in their own orbits at any one time). The nineteenth-century historian Ignatius Donnelly tells the story this way:

> On the 27th day of February 1826, M. Biela, an Austrian officer . . . discovered a comet in the constellation of Aries, which, at that time, was seen as a small, round speck of filmy cloud. Its course was watched during the following month by M. Gambart at Marseilles and by M. Clausen at Altona, and those observers assigned to it an elliptical orbit with a period of *six years and three quarters* for its revolution.
>
> M. Damoiseau subsequently calculated its path, and announced that on its next return the comet would cross the orbit of Earth, *within twenty-thousand miles of its track,* but about *one month before the Earth would have arrived at the same spot!*
>
> This was shooting close to the bull's-eye!
>
> He estimated that it would lose nearly ten days on its return trip, through the retarding influence of Jupiter and Saturn; but if it lost forty days instead of ten, what then?
>
> But the comet came up to time in 1832, and the Earth *missed it by one month.*
>
> And it returned in like fashion in 1839 and 1846. But here a surprising thing occurred. *Its proximity to Earth had split it in two;* each half had a head and a tail of its own; each had set up a separate government for itself; and they were whirling through space, side by side, like a couple of racehorses, about 16,000 miles apart, or about twice as wide apart as the diameter of Earth.
>
> In 1852, 1859, and 1866, the comet SHOULD have returned, but it did not. It was lost. It was dissipated. Its material was hanging around Earth in fragments somewhere.[74]

On the last occasion, 1866, another commentator tells us that "in November, the period of Biela's return, the world beheld a most brilliant meteor shower, and in 1872, 1885, and 1892, corresponding with its former orbit, there were imposing displays of meteors in November.[75] At one site more than 160,000 shooting stars were seen in an hour and even today the debris of Comet Biela returns annually as the Andromedid meteor shower.[76]

On its way into the inner solar system the Great Comet of 1744 transformed itself near the orbit of Mars into six large, luminous fragments each with its own tail from 30 to 44 degrees in length.[77] On 4 October 1994, Jim Scotti of Spacewatch reported that comet Harrington—which does not cross the orbit of the Earth—had broken into at least three parts.[78] In March 1976 the nucleus of comet West disintegrated into four parts.[79] And we have seen how comet Shoemaker-Levy 9 broke into 21 fragments.[80]

Other examples of fragmentation include comet Macholz 2, which was found by the astronomer Donald Macholz in 1994 in a region of the sky not yet covered by any of the telescopes of the world's skeletal Spacewatch network.[81] This comet is on an Earth-crossing orbit with a short period of about seven years and consists of a swarm of six individual nuclei still relatively close to one another but drifting apart—indicating that they were probably produced by the fragmentation of an original larger nucleus sometime in the 1980s.[82]

The remarkable Kreutz "sun-grazing" comets—so bright that they have sometimes been seen in daylight—are a similar family of nuclei descended from a common progenitor. Consisting now of about a dozen individual objects on virtually identical orbits but with varying periods—from 500 to 1,000 years—they pass very close to the surface of the Sun, some to within just half a million kilometers of its surface.[83] Indeed in 1979 one of these comets crashed directly into the Sun, being photographed just before it did so by the U.S. Navy satellite *Solwind.* The impact caused "a brightening over half the solar disk, which lasted a full day."[84]

Tracing back the orbits of the Kreutz sun-grazers, Victor Clube and Bill Napier conclude:

> They were once a single, gigantic object, ten or twenty thousand years ago, which underwent a hierarchy of disintegrations. There is little doubt that the tidal strain induced by the close passage to the Sun has split the parent comet into fragments.[85]

We saw the effects that such fragments can have when comet S-L 9 crashed into Jupiter.[86] Since any lesser planet would have been killed by those 21 hurtling projectiles, we are led to wonder whether it might not have been precisely such an incident—although perhaps on an even grander scale—that killed Mars?

Could a gigantic comet be implicated in the dark story of the Martian past and also, perhaps, in the uncertain future of Earth?

Voyager on the Abyss

FROM the very beginning of their great civilization the ancient Egyptians conceived of the mission and predicament of mankind as being inseparably connected to the cosmos and governed by it. They were certain that our true spiritual home was in the heavens, from whence we descended only temporarily into the material world, and that "the inhabitants of heaven" exercise a powerful influence upon our lives, which we neglect at our peril. In their teachings the stars and the planets were gods, not just remote points of light in the sky, and meteorites made of *bja* iron—the "divine metal"—represented an interchange between the spiritual and material realms.

Such ideas were present from the earliest historical period and are expressed in the Pyramid Texts, the oldest-surviving scriptures of mankind. Together with the later funerary literature of the ancient Egyptians, they teach that a secret path of pure *knowledge* exists—"a way of ascent to the sky"[1]—that can lead us back to our heavenly home if we search it out and make ourselves masters of it. Nor can there be any doubt that the ultimate goal of the ancient Egyptian initiates was a form of conscious immortality—the "life of millions of years"—which would be achieved through rebirth as a star:

> O King, you are this great star, the companion of Orion who traverses the sky with Orion, who navigates the Duat with Osiris. You ascend from the east of the sky, being renewed at your due season and rejuvenated at your due time. The sky has borne you with Orion.[2]

The reader will recall that the Duat sky region—the ancient Egyptian netherworld, a starry afterlife kingdom—was dominated by the constellations of Orion, Taurus, and Leo and divided by the "Winding Waterway," which we call the Milky Way:

> The celestial portal to the horizon is opened to you, and the gods are joyful at meeting you. They take you to the sky with your soul . . . You have traversed the Winding Waterway as a star crossing the sea. The Duat has grasped your hand at the place where Orion is, the Bull of the Sky [Taurus] has given you his hand.[3]

The Milky Way is our galaxy and the great sky river that we see is made by the combined light of billions of stars lying along the plane of the galactic disk.[4] Within the galaxy, which is technically a "spiral galaxy,"[5] all stars are indeed in motion, sailing across the Catherine wheel of spiral arms, orbiting the galactic nucleus. Our particular star, the Sun, has recently passed through the Orion spiral arm,[6] so named because it contains the spectacular Orion nebula, which lies beneath the three belt stars of the constellation of Orion. Astronomers have put forward intriguing evidence that the passage was a "bumpy" one, that the solar system was severely disturbed by it and that the consequences of this disturbance have included a series of spectacular sky events during the past 20,000 years—all of them seeming to emerge from the constellation of Taurus.[7]

SKY/GROUND MESSAGE

It may not be a coincidence that the ancient Egyptians had a deep and abiding interest in the constellations of Orion and Taurus. Their belief that this area of the sky is the cosmic home to which we should strive to return is expressed not only in religious texts but also in the three great pyramids of Giza and in the so-called Bent and Red pyramids of Dashur.[8] Standing at the geodetically significant location of 30 degrees north latitude (one-third of the way between the equator and the North Pole) and incorporating a series of mathematical constants, transcendental numbers (i.e., numbers not capable of extension in terms of a finite number of arithmetical operations), and geometrical ratios such as *phi*, *pi*, and *e/pi*, the Giza group mimic the sky image of the belt stars of Orion while the Dashur pyramids mimic the relative positions of two stars in the constellation of Taurus—Aldebaran and epsilon Tauri.[9] It is likely that the Red pyramid—representing Aldebaran—was built of red stone because of the conspicuous color of its stellar counterpart, which forms "the glinting red eye" of the Taurus sky bull.[10]

We showed in chapter 17 that precisely the same logic is expressed in the enigmatic figure of the Sphinx—painted red because of its association with the Red Planet Mars and lion-bodied to mimic the sky image of the constellation of Leo rising at the spring equinox. No civilization that understands precession should have any more difficulty than ourselves in working out that Leo last "ruled" the equinox between approximately 13,000 years ago and 10,000 years ago. We are sure that the builders of the Sphinx intended this connection to be made. This is why we wonder if it is possible that part of the "message" of the Sphinx may simply be: "consider Mars when the spring equinox was in Leo."

The fact is that when we do consider Mars we find the following:

- It once had rainfall and running water and could have supported life. We do not know when this was. There are some indications that it could have been extremely recently.
- It has upon its surface an object that looks very much like the face of a Sphinx set among a conglomeration of other objects including several that greatly resemble pyramids. We have seen that these Martian "structures" are set at a geodetically significant latitude and incorporate many of the same mathematical properties as the monuments of the Giza necropolis.
- The Martian surface has been devastated by collisions with a gigantic swarm of cosmic debris—including three huge world-killing projectiles up to several hundred kilometers in diameter that caused the Hellas, Argyre, and Isidis craters. We saw in part 1 that this cataclysm need not necessarily have happened in some remote geological period, as scientists have tended to assume, but could have occurred quite recently, perhaps less than 20,000 years ago—perhaps even in the same period in which Earth's last Ice Age was suddenly and mysteriously ending amid planet-wide extinctions of animal species.[11]

Is it possible, in other words, that "the terminal Mars cataclysm" and the lesser but still very severe cataclysm that brought Earth out of the last Ice Age could both have occurred *at more or less the same time*—and perhaps even have been caused by the same agent?

If we think as the ancient Egyptians did, seeing the cosmos, the earth, the planets, and all the stars as the constituent parts of a continuous interconnected matrix, then we will find it easier to understand what modern science has only recently proven to be true—namely that the solar system and all the planets are profoundly influenced by the galaxy and that these influences flow in toward us from deep space like tides.

THE JOURNEYS OF RA

The ancient Egyptians depicted the Sun—the god Ra—as a voyager upon the waters of the abyss:

> Men praise thee in thy name of Ra. . . . Millions of years have gone over the world; I cannot tell the number of those through which thou hast passed. . . . Thou dost pass over and dost travel through untold spaces requiring millions and hundreds of thousands of years to pass over. . . . Thou steerest thy way across the watery abyss to the place which thou lovest . . . and then thou dost sink down and make an end of hours.[12]

Although the text is from the Book of the Dead, the ideas it expresses are the territory of modern astrophysicists, who have learned that everything in the universe is in motion and that as the Sun makes its way around the galactic nucleus, it is indeed a traveler through "untold spaces" that require "millions of years to pass over."

In fact, a number of different motions are involved. Here are the basics:

(1) Drawing with it the entire solar system, including of course all the comets of the Oort cloud and the Kuiper belt, the Sun is locked in a vast orbit around the galactic nucleus, completing each revolution in a period of approximately 250 million years.[13] Traveling at a speed of 225 kilometers per second, it has recently passed through the Orion spiral arm on the inner edge of which it now stands.[14]

(2) The Sun orbits the galactic nucleus faster than some stars and slower than others—in general stars distant from the nucleus travel at lower speeds than those closer to it, and the Sun is located relatively far from the nucleus.[15] "It's a complete muddle," explains Victor Clube:

> Everything passes through everything else. I mean a star doesn't pass through another star. But space in general is so empty that all these features that we talk about sort of interpenetrate. . . . So the Sun is actually moving in its particular orbit. And it happens to be going at a different speed from any old spiral arm or any old molecular cloud. So it passes through these things.[16]

(3) The Sun does not always travel in the flattish (although light years' thick) horizontal plane of the galactic disk. Instead it's motion is better understood as wave-like (astronomers have compared it to the motion of a carousel horse,[17] or a porpoise[18]). The effect of this slow undulation is that the Sun in its orbit periodically swims up above the dense central plane of the galaxy, then dives down again into it, then emerges beneath it, then swims up once more—and so on, endlessly, as it pursues its circuit. The rhythm of these movements is regular and cyclical with the Sun rising from its lowest point beneath the disk to its highest point above it in a period of just over 60 million years and falling again to the lowest point after a further 60 million years. It is only at the halfway points in this journey, therefore—roughly every 30 million years—that it passes through the galaxy's dense central plane.[19]

(4) Superimposed on the Sun's predominantly circular (albeit up-and-down) trajectory about the galactic nucleus there is also what astronomers refer to as the "peculiar" solar velocity.[20] According to the calculations of Mark Bailey, Victor Clube, and Bill Napier:

> This may be represented as a vector directed respectively toward the galactic center, parallel to the circular velocity and perpendicular to the galactic plane. In galactic coordinates this corresponds to a motion toward [a point] roughly 30 degrees out of the plane toward the north galactic pole. This direction, incidentally, can be viewed from the northern hemisphere on any summer's evening, as it lies . . . roughly halfway between the bright stars Vega and Ras Alhague, *almost exactly opposite the molecular clouds in Orion* [author's emphasis].[21]

We remind the reader that the pyramids of Giza, which model the belt stars of Orion, are located at 30 degrees north latitude on Earth—or, to put it another way, at a point "roughly 30 degrees out of the plane of the equator toward the north geographical pole." Moreover this place in the galaxy toward which the Sun is vectored ("thou steerest thy way across the watery abyss to the place which thou lovest . . . and then thou dost sink down and make an end of hours") is located opposite the molecular clouds of the Orion nebula. As the Hubble space telescope conclusively demonstrated during the 1990s the nebula is a star-forming region—literally a place where new stars are being born.[22] Lying in a region of space through which the Sun and Earth are estimated to have passed roughly 5 to 10 million years ago,[23] it forms the feature of the Orion constellation, beneath the belt stars, which the Greeks depicted as a sword but the ancient Egyptians saw as the phallus of Osiris, the god of rebirth.

AS ABOVE, SO BELOW

The ancient Egyptians believed that events on Earth are governed, conditioned, and directly affected by events in the sky and that "all the world which lies below" is

> set in order and filled with contents by the things which are placed above; for the things below have not the power to set in order the world above. The weaker mysteries, then, must yield to the stronger . . . the system of things on high is stronger than the things below . . . and there is nothing that has not come down from above.[24]

This is literally true of comets. Not only do they "come down from above" in the sense of belonging to the sky, occasionally colliding with planets, but they are also, as astronomers now know, periodically propelled toward the inner solar system by even more distant forces at the level of the galaxy. Such influences from "on high" are governed largely by the character of the different deep-space environments that the Sun encounters as it pursues its immense circular and undulating course around the galactic nucleus and are felt most strongly during passages through the galaxy's dense central plane.[25]

Two key factors are involved, both of which, in reality, interpenetrate. These are the galactic "spiral arms" and the massive nebulae—found often but not exclusively within spiral arms—that are known as gigantic molecular clouds.

COMET FACTORIES

A degree of controversy exists among astronomers as to what spiral arms actually consist of, but most would agree with Victor Clube that they are relatively transient features, ejected from the galactic nucleus, and that the galaxy is constantly generating new ones: "So it kind of grows leaves, seasonally, if I can put it that way. . . . I see lots of comets condensing out of the hot gas that's originally in spiral arms. And it's these comets which aggregate to make the stars."[26]

We are reminded of electrifying spectroscopic evidence reported by the astronomer Lagrange-Henri in 1988, of "a swarm of small cometary-like bodies falling at high velocities toward Beta Pictoris, a relatively young star around which planet formation is either occurring now or has just been completed."[27]

Condensing in the hot gas of spiral arms, such comets may reach gigantic sizes. Clube and Napier report that truly massive examples have been identified "in the vicinity of two well studied and exceedingly active stellar associations, namely the so-called Gum Nebula and the Orion Nebula."[28] These comets are

> vast compared to solar system examples, the tails being up to a million times longer. . . . The tails are not only pointing away from the center of the parent association where most of the local radiation originates, but the heads seem to be in highly eccentric orbits moving away from the central source. . . . It is supposed that the heads may comprise huge assemblages of interstellar comets or planetesimals. . . . We thus have an indication that we may be dealing here with large, loose aggregates of cometary material which are either about to be or are in the process of forming new stars.[29]

As well as being the nurseries of gigantic interstellar comets, spiral arms are thought to contain a mass of other material varying in size from the tiniest gas and dust particles up to objects "as big as the moon"[30]:

> The galactic evidence favors spiral arms containing planetesimals or comets in all their variety of forms. It is inevitable then that the solar system interacts with such material as it passes through the spiral arms.[31]

The Sun can take anywhere from 50 million years to 100 million years to make a complete horizontal passage across a spiral arm.[32] Since spiral arms tend to be located at or very near the galactic plane,[33] the Sun's porpoise-like up-and-down motion means that it will spend most of its time either above or below the arm, only diving into the arm itself at cyclic intervals of approximately 30 million years.[34]

MONSTER CLOUDS

The second periodic "hazard of the galactic plane"—the flattened zone where most loose cosmic material tends to gravitate—is the possibility of encounters with gigantic molecular clouds (GMCs). As noted, these can be found as complicating factors within already "lumpy" spiral arms, or can exist in isolation, lying in the interstellar medium between spiral arms.

GMCs are typically about 100 light years across and have a mass (as distinct from diameter) estimated to be about half a million times that of the Sun.[35] The basic matrix of these cold, massive concentrations consists of molecules of hydrogen gas and more complex compounds, mixed with dust.[36] In addition, they often contain dense concentrations of young stars and, Clube and Napier believe, "enormous numbers of newly formed comets as well . . . circulating freely within the nebula."[37]

"Confined within the flat plane of the Milky Way," it is estimated that "a few thousand" GMCs orbit the galaxy.[38] Inevitably, there will come times, again governed by the 30-million-year periodicity with which the Sun's own orbit oscillates in and out of the galactic plane, when it must penetrate GMCs:

> Close encounters between the Sun and such nebulae, say to within a few light years, have probably occurred more than fifty times during the lifetime of the solar system. Actual penetration has probably occurred more than a dozen times, several involving passage of the Sun to within about a light year of the cloud center.[39]

GALACTIC CONTROL

We now have all the pieces in place to understand that comets find their way into the inner solar system, and can threaten the destruction of worlds, not because of some nearby "local" event but because of the distant and almost unimaginable influence of the galaxy. In other words, in the purest sense, what happens down here "below," on Earth—or on Mars—when a comet approaches closely, can indeed be traced back far "above" to the cycles of the cosmos.

Astronomers have shown that passage through a GMC has a profoundly destabilizing effect on the Oort cloud (the hollow sphere of 100 billion comets that surrounds the outer reaches of the solar system) and that occasional passages past exceptionally dense, concentrated "substructure" within the GMCs has a "relatively more damaging effect."[40] At one and the same time, the GMC "strips away" the outer layer of the shell of comets and carries it off while its immense gravitational tides propel other comets inward toward the Sun.[41] Embarking on a journey that will take millions of years to complete, these fallen angels gradually spiral down through remote space. Some enter a kind of limbo in the Kuiper belt where they may remain for as much as 3 million years before beginning to fall in again toward the center. Others take a more direct route and eventually find themselves within the gravitational influence of one of the giant planets, which whirls them around like pinballs and projects them on new courses toward the inner solar system.[42]

Passage through a spiral arm has equally dramatic effects. Here the Oort cloud is replenished with new interstellar comets and other "large, solid bodies" that have grown in the spiral arm.[43] Indeed, it is estimated that "the solar system, acting as a gravitational scoop, captures billions of such bodies when it crosses spiral arms."[44] As these bodies swarm into the Oort cloud they propel other comets out of the cloud and toward the Sun, leading to increased cometary activity in the inner solar system.[45] Eventually "episodes of planetary bombardment occur,"[46] sustained over long periods with "profound biological and other consequences."[47] At each episode, huge quantities of material are unleashed within the solar system, representing a lingering threat that can strike any time, or repeatedly, over many thousands of years.

In both cases—GMCs and spiral arms—the cycle of disturbance that leads to the planetary bombardments is primarily governed by the porpoise-like up-and-down motion that takes the Sun through the galaxy's dense central plane at intervals of about 30 million years. Astronomers also recognize a second, longer rhythm at work—a cycle of around 250 million years, linked to the period of the Sun's orbit around the galactic nucleus.[48]

In other words, the entire comet flux into the inner solar system is controlled at the galactic level, and the comets themselves represent fragments of the galaxy flung upon the planets. During severe encounters with GMCs, or particularly bumpy spiral-arm passages, it is to be expected that *waves of potential impactors, some of them in the world-killing 200-kilometer-plus range,* will be released to work their way down toward the Mars-Earth-Moon realm; these waves, moreover, will follow earlier waves released by previous galactic encounters and will be followed by further waves from future galactic encounters.[49] The inner planets, in other words, will continue to face periodic bombardments that we may expect to be both heavy and sustained. So long as the Sun still shines and comets continue to be manufactured in spiral arms, the process can go on forever.

PULSE

The heartbeat of the process is that pounding 30-million-year cycle—modulated by a 250-million-year cycle—produced by the Sun's oscillations through the galactic plane. As a result of tenacious detective work, multidisciplinary teams of scientists, including astrophysicists, astronomers, mathematicians, geologists, and paleontologists, have been able to establish *a close statistical correlation* between these great comet-multiplying cycles of galactic disturbance, the dates of known craters on Earth, and mass extinctions of animal species[50]

> with major extinctions occurring every 250 million years or so, due to the passage of the solar system through a spiral arm of the galaxy, and lesser extinctions occurring approximately every 30 million years as the solar system crosses the galactic plane. . . . The fact that interstellar clouds are not all found exactly on the mid-plane of the galaxy would explain why not all extinctions seemed to have occurred precisely on schedule, the standard deviation of each individual episode being 9 million years.[51]

Sir Fred Hoyle and Professor Chandra Wickramasinghe of the University of Cardiff have firm opinions about the K/T object that caused the extinction of the dinosaurs 65 million years ago:

> The evidence is that a giant comet plummeted into the inner solar system, passing close enough to Jupiter to fragment it into many pieces approximately 65.05 million years ago. Repeated passages by Jupiter over a 100,000-year period produced hierarchical fragmentation, and one such fragment (of normal comet size) came close enough to Earth to crash onto the planet's surface.[52]

As Hoyle and Wickramasinghe also point out, the mass extinction of 65 million years ago was not an isolated incident but part of a cycle that is hard to miss during the past 100 million years, with mass extinctions at 94.5 million years ago, 65 million years ago, and 36.9 million years ago.[53] The sediments of these epochs "have been found to be associated with iridium enhancements, so a cometary connection is believed to follow."[54] In addition, studies of impact craters on Earth and of crater samples brought back from the Moon show that intense, sustained, and violent bombardments have taken place with approximately the same periodicity.[55]

Within the margins of tolerance this data warns us that the Earth-Moon system could now enter an episode of bombardment at any time. Indeed, as we will see in the next chapter, an increasingly large and eminent group of scientists believe that we have been in such an episode for almost 20,000 years, that it is implicated in the sudden and mysterious end of the last Ice Age—which resulted in mass extinctions and a global flood—and that the worst is yet to come.

What no one has considered, perhaps because it seems so far away when viewed from Earth, is the haunting possibility that Mars, which the ancient Egyptians called Horus the Red, and the Aztecs Xipe-Xolotl, the "Flayed Planet,"[56] could also have been a victim of that same sustained bombardment.

24

Visitor from the Stars

THE mystery of what happened to Mars is a jigsaw puzzle that has been scattered throughout the galaxy—perhaps even beyond the galaxy—and across billions of years. Moreover, since the distance between Mars and Earth is insignificant on the galactic scale, it is reasonable to suppose that any influence felt by Mars will also have been felt by Earth—and vice versa. The picture that has begun to emerge sets the solar system within its galactic environment and shows us that *a clear and present danger is posed by comets.*

The danger is as yet extremely difficult to quantify, and because of this the precise risks are impossible to assess. All that we know for sure is that as the Sun orbits the galactic nucleus, towing the Oort cloud, the Kuiper belt, Mars, Earth, and all the other planets in its wake, it exposes *every one of them* to periodic surges of cometary activity whenever it passes through a spiral arm or a gigantic molecular cloud. As though propelled by some great cosmic tide, waves of comets are unleashed by such encounters and roll toward the inner solar system—including, at random intervals, giant comets hundreds of kilometers across.

It may take the missiles in each wave *millions* of years to fall far enough to cross the orbits, and enter the domain, of the stony planets. During this long spiraling-down process, in which comets have their own orbits repeatedly "nudged" and stressed by interactions with the gas-giants Neptune, Saturn, and Jupiter, many are ripped apart by gravitational forces and split up into multiple fragments—thus vastly increasing the total numbers of projectiles.

We will argue that much of the damage done to Mars, and such enigmas as the planet's strange crustal dichotomy, may be accounted for by a single head-on collision with the fragments of a truly gigantic comet that came in from the outer solar system on such a wave. Moreover, when we look at the ruined, cratered corpse of Mars, so grim and dead, so tragic with its empty rivers and its dry oceans, is it not obvious that worlds *can* be killed by comets? And is it not obvious, too, as the old saw goes, that there but for the grace of God go you or I?

CYCLES OF THE HEAVENS

Science has not yet been able to bring back any samples from Martian craters or to undertake a detailed geological investigation of the planet. Almost all of our assumptions about Mars are based on what can be learned from studying photographs taken by orbiting spacecraft—and these cannot tell us *when* the terminal Mars cataclysm occurred. As we have maintained throughout this book, the thousands of impact craters south of the line of dichotomy need not have accumulated slowly, over billions of years, as most scientists still believe, but could have been inflicted *suddenly,* perhaps even in one single cataclysmic incident, and perhaps *recently.*

This is a hypothesis that can be tested when manned landings are made on Mars. Until then it is only an assumption, and definitely not a proven fact, that the Martian craters are billions of years old. A certain light may be shed on the matter, however, by what we know for sure has happened on Mars's nearby neighbor, Earth. Here we do not need to rely on grainy photographs taken by orbiters thousands of kilometers up but can look into tangible and empirical matters such as extinction records, data gathered from craters around the world, chemical tests on soil samples—and so on and so forth.

What these indicate, as we reported at the end of the preceding chapter, is that our planet has experienced cyclic episodes of bombardment and extinction at regular intervals during the past 100 million years—specifically 94.5 million years ago, 65 million years ago (the K/T event), and 36.9 million years ago.[1] We have also shown that the cycle has a basic "heartbeat" of 30 million years with "the standard deviation of each individual episode being 9 million years."[2] In plain English this means that if you look at the cycle over a long enough period of time—several hundred million years—you will find that linked bombardment and extinction episodes do occur at roughly 30-million-year intervals, but that the gap may become as small as 21 million years in some cases, or as large as 39 million years in others.

Returning to the last 100 million years we find that the intervals between extinction events have been consistently within this range. Between 94.5 million years ago and 65 million years ago the figure works out at 29.5 million years. Between 65 million years ago and 36.9 million years ago the figure works out at 28.1 million years. Since we know that the bombardments are caused by waves of galactic material that swamp the entire solar system—not just near-Earth space—we think it is a good guess that Mars, and the Moon, would have experienced bombardment episodes, pretty much in tandem with Earth, at around 94.5, 65, and 36.9 million years ago. As we saw in the last chapter this has already been confirmed in the case of the Moon. In the case of Mars it is another testable hypothesis that will have to await a manned landing—but then so will all hypotheses about Mars, from all sources. For neither the wild theories of the craziest cranks nor the sober reflections of celebrated scientists have yet had to be proven against hard empirical evidence from the surface of the planet itself.

To reiterate, it is our hypothesis that Mars and Earth both experienced bombardment episodes at around 94.5, 65, and 36.9 million years ago. The final interval, from 36.9 million years ago up till today, is significantly longer then the previous two. Indeed, it is dangerously close to the extreme upper limit of the cycle—39 million years.

Could we be nearing the end of what is already beginning to look like an untypical and overlong period of quiescence? Could another bombardment of the inner planets be on the way?

WHERE ARE WE NOW?

The first steps toward an intelligent assessment of our current predicament have already been taken by a group of leading astronomers including Victor Clube and Bill Napier, David Asher, Duncan Steel, Mark

Bailey, Sir Fred Hoyle, and Professor Chandra Wickramasinghe. There is not space here to report all of their discoveries, so in the rest of this chapter we will inevitably have to focus on the central evidence chain that they have built up. We shall do so as far as possible in their own words, which convey to the reader better than we can the deep concern and rising sense of urgency that these scientists feel. We share their concern. And we believe it is a matter of fundamental importance that the public and policymakers should be made aware of their work—which demonstrates that the galactic environment in which the solar system presently finds itself is a uniquely deadly and unpredictable one. Together with a growing number of colleagues from many other countries, they draw particular attention to the following facts:

1. There is evidence of "a very recent disturbance of the Oort cloud related in some way to the solar motion."[3]
2. The Sun has recently passed through the galaxy's densely crowded mid-plane and is presently "skimming" just 8 degrees above it.[4]
3. For the past 100 million years or so the Sun has been visiting the Orion spiral arm,[5] crossing it "at a fairly narrow angle to the axis, completing one or two porpoise-like cycles as it does so."[6]
4. The Sun has recently completed said passage above and is now poised just above the inner edge of the arm.[7]
5. Here the Sun has "penetrated what appears to be the remains of an old, disintegrating giant molecular cloud. This is a ring of material which incorporates most of the molecular clouds and star-forming regions in the solar neighborhood. The young blue stars form an arc in the sky now known as Gould's belt but recognized since the time of Ptolemy. . . . The solar system passed through Gould's belt only 5–10 million years ago."[8]
6. The chilling conclusion is that the Sun's current "address" in the galaxy not only indicates that a bombardment episode is imminent but that *it must have already begun* and that the impact rate *at the present time* should be exceptionally high:

> The Sun's position at the inner edge of the Orion spiral arm ensures that we are currently in an active phase. Further, the solar system has just passed through the plane of the galaxy where the tidal stresses acting on

> the comet cloud are at their maximum; the comet flux is therefore near a strong peak of its galactic cycle. It has also recently passed through Gould's belt and is therefore undergoing an exceptional tidal stress due to a recent passage through an old, disintegrating molecular cloud. . . . This encounter must have created a sharp impact episode, within which we are still immersed. . . . [Indeed] the conditions which would yield an exceptional flux of comets on to Earth—positioning near the galactic plane, proximity to a spiral arm, and recent passage through a system of molecular clouds—are all simultaneously met by the solar system at the present time. . . . We are in an impact episode now.[9]

THE TRAIL OF A GIANT COMET

The detective work that the astronomers have done pinpoints the Sun's turbulent passage through Gould's belt as the single most likely source of the episode. Near the end of the passage, around 5 million years ago, they believe that a wave of comets was expelled from the Oort cloud by tidal stress and began the slow, light-year-long journey toward the inner solar system. Among these comets was at least one giant "up to a few hundred kilometers in size"[10] that took several million years to spiral down toward the planets. There it first entered the realms of Neptune, Saturn, and Jupiter, where it was detained for perhaps another million years as its orbit gradually decreased in size while at the same time evolving into an ever more elliptical form. As recently as 50,000 years ago, a gravitational "kick" from Jupiter finally brought it into the inner solar system, where it settled into a steeply elliptical orbit with a perihelion very close to the Sun and an aphelion just beyond Jupiter.[11] Such an orbit would inevitably be both Earth-crossing and Mars-crossing. Victor Clube told us:

> We have a very specific picture that this giant comet was deflected into a sun-grazing orbit. Now that's one that goes very close to the sun. And also highly eccentric, meaning that it gets very close to Jupiter as well. Now this very narrow, elliptical orbit is the key to the evolution of this particular giant comet. The frequent passages close to the Sun ultimately cause the comet to break up into lots of lumps. But it doesn't do it straight away. This is a long-drawn-out process.[12]

The process did not begin in earnest until about 20,000 years ago—although some of the astronomers suspect it could have been as recently as 15,000 to 16,000 years ago,[13] when a major change seems to have overtaken the giant comet.[14] The approximate date of this event has been established by dynamical studies, and from samples of interplanetary dust taken from Earth and the Moon (which show that a great flux occurred between 20,000 and 16,000 years ago[15]) and is likely to be correct, give or take a couple thousand years.[16] Astronomers, however, are much less certain about exactly *what* happened in that crucial epoch.

One possible line of speculation is that the original object had become so volatile as a result of repeated passes close to the Sun that it literally tore itself apart in an explosive fragmentation. Another, perhaps more plausible, is that it trespassed the Roche limit of a planet—as comet Shoemaker-Levy 9 did during 1992–1994—and was pulled to pieces by intolerable tidal stresses.[17]

This is a puzzle to which we will need to return.

MILLIONS OF PIECES, THOUSANDS OF YEARS

Whatever the precise nature of the original fragmentation event, astronomers have demonstrated that it was followed by a very lengthy and continuing "hierarchy of disintegrations" spread out all along the path of the comet's orbit and periodically bombarding all the inner planets with dense meteor streams, fireballs, and short-lived swarms of Tunguska-sized projectiles, together with

> many individual asteroids of a kilometer in size or greater, which themselves break up, and at least one conspicuously large core remnant which is probably enveloped in a swarm of dust and debris.[18]

Sir Fred Hoyle points out that when the original giant comet was still in its undivided state the chances of a collision with Earth were small—he estimates only about one part in a billion on each orbit:[19]

> But as [such a] comet divides into more and more chunks the chance of one or another of them hitting Earth rises inexorably, until one or another of them will indeed score a bull's-eye on our planet.[20]

Within 10,000 years of the initial explosive fragmentation event, Hoyle estimates that the original comet would already have "divided into about a million pieces" with an average weight in the range of 10,000 million tons each (implying a weight of 10,000 *million million* tons for the mother object).[21] Further hierarchical disintegrations into smaller and smaller—and more and more numerous—pieces would then have followed, spread out over an immensely long period, with the rate of individual collisions rising as the numbers of available projectiles increased.[22]

It is obviously important to know how long such a process might be expected to continue.

Victor Clube calculates that the "comminution lifetime" of a giant comet after fragmentation begins—that is, the time it will take to reduce itself to pieces too small to cause impact damage—may be as long as 100,000 years.[23] Since the first major fragmentation event of the comet that we are interested in is thought to have occurred only 20,000 years ago, the implication is that swarms of deadly projectiles of assorted sizes are still likely to be orbiting along the Earth-crossing path previously pursued by the original intact comet.[24] Moreover there is a chilling possibility that the larger nuclei remaining in the swarm could prove extremely difficult for astronomers to detect "due to their immersion in obscuring dust—giving them overall something of the character of a 'holy grail.' "[25]

The laws of probability suggest that if such a near-invisible menace is indeed lurking on an Earth-crossing orbit, then fragments from it should have collided with the Earth-Moon system several times during the past 20,000 years.

HIDDEN HAND

Clube, Napier, Hoyle, Wickramasinghe, and their colleagues have demonstrated that precisely such a series of encounters may have been the hidden hand at work behind the sudden, catastrophic, and hitherto unexplained end of Earth's last Ice Age.[26] This meltdown began 17,000 years ago, reached two dramatic peaks at around 13,000 and 10,000 years ago, and by 9,000 years ago had freed the world of ice sheets that had been stable for the previous 100,000 years.[27]

This immense and—in geological terms—extremely rapid change is one of the central mysteries explored in *Fingerprints of the Gods,* which further argues that the cataclysm that ended the last Ice Age also obliterated almost all traces of a highly advanced prehistoric civilization. It is our hypothesis, explored now in a number of works, that there were survivors of that lost antediluvian civilization (a global flood with tidal waves hundreds of meters high was one of the most devastating consequences of the terminal Ice Age cataclysm), and that they spread out all around the world, passing down myths and traditions of a golden age brought cruelly to an end—the biblical story of the Flood of Noah is a classic example. We are also firmly of the opinion that something more than myths and traditions has been preserved from "before the flood"—even to this day—in initiation teachings transmitted by secretive groups and in certain compelling works of architecture, *of unestablished provenance,* such as Stonehenge in England, Teotihuacan in Mexico, and the pyramids and the Great Sphinx of Giza.[28]

Occurring as they do on a devastated planet that has indisputably suffered a grand impact cataclysm that caused (among other effects) gigantic floods and tidal waves kilometers high, the reader will appreciate why we could not turn our backs on the enigma of the pyramids and Sphinx-like Face of Mars—whatever they may ultimately be proved to be.

Parallel worlds?

Parallel cataclysms?

Parallel lost civilizations?

Who knows? Some mysteries, surely, are worth looking into just because they are there—even if final answers may never be forthcoming.

What is certain, meanwhile, is that the inner solar system has experienced a great surge of cometary activity in the past 20,000 years, that Earth has suffered a mysterious cataclysm during this period and that Mars has also suffered a mysterious cataclysm (although there is as yet no proof as to when). These traumas were severe enough in the case of Mars to snuff out the planet *altogether* as a habitat for life, and in the case of Earth to cause the extinction of an estimated 70 percent of species and to raise sea levels by more than 100 meters.[29]

GRAVE CONSENSUS

We need not repeat here the evidence and arguments, already fully developed by ourselves and others, in *Fingerprints of the Gods* and elsewhere, concerning the spectacular disaster that shook the earth at the end of the last Ice Age. But the great challenge with which this evidence confronts researchers is the need to work out what sort of event could possibly have caused such a *massive* disaster on such an astonishing worldwide scale. Lengthy consideration was given in *Fingerprints* to Charles Hapgood's theory of crustal displacement—which was then being strongly advocated by the Flem-Aths in Canada[30]—but very little attention was paid to the possible role of cosmic impacts, either as trigger factors in displacements (see discussion in chapter 18) or as direct causative agents.

We were not alone in this oversight. Throughout most of the twentieth century, Western science as a whole has resolutely ignored the role of impacts in Earth history, only gradually and reluctantly waking up to their significance in the light of the irrefutable evidence of a cometary collision at the K/T boundary (not fully accepted until 1990) and such dramatic events as comet S-L 9's breakup into 21 fragments and subsequent bombardment of Jupiter in 1994. When the fragments struck, humanity was offered a glimpse through the gates of hell. Since then, after being almost dismissively ignored for two decades, the theories of catastrophist astronomers such as Clube, Napier, Hoyle, and Wickramasinghe have achieved rapid acceptance among the vast majority of their peers.[31]

Fingerprints of the Gods went to press early in 1995. During the lengthy investigation that underlies *The Mars Mystery* we became aware of the growing catastrophist consensus within astronomy. It is a grave consensus, involving many eminent scholars, and it has profound implications that have not yet been properly communicated to the public. We find ourselves today in more or less complete agreement with this new consensus, which holds, as Clube and Napier put it,

> that great impacts, occurring within bombardment episodes as the solar system moves through spiral arms, have been a major controlling factor in the evolution of life, being responsible for catastrophic mass extinctions of species. Fundamental geological phenomena such as frequent sea-level changes, the occurrence of Ice Ages and plate tectonic episodes, including mountain building, may also have been triggered by impacts.[32]

More specifically, although we do not rule out a crustal displacement as a complicating factor in the terminal Ice Age cataclysm that took place between approximately 17,000 and and 9,000 years ago, we are now persuaded that the astronomical theory of impacts connected to the decay and fragmentation of a giant comet provides not only the most plausible but also the clearest and simplest explanation for all the events and enigmas of those crucial 8,000 years.[33] Because this was precisely the period in which humanity emerged from the darkness of the Ice Age and onto the threshold of modern history, and because, as we will see, there have been other impacts much more recent than 8,000 years ago, we agree with Hoyle and Wickramasinghe:

> The history of human civilization bears witness to the most recent chapter in a series of cosmic events that controlled our planet in a decisive way.[34]

THE TESTIMONY OF BEETLES

Looking into the geological record, and such arcane matters as the carcasses of temperature-sensitive beetles (the presence or absence of particular species in given strata provides a precise temperature chart for the epochs in which those strata were laid down[35]), Hoyle and Wickramasinghe have produced a revealing chronology of key Ice Age events.

They have shown that although melting of the ice sheets did begin at around 17,000 years ago, proceeding sporadically from there in a series of advances and retreats—perhaps as a result of a parallel series of small impacts—the most spectacular temperature rises occurred in two isolated incidents, one somewhere between 13,000 and 12,000 years ago and the other somewhere between 11,000 and 10,000 years ago.[36]

Here is Fred Hoyle's account of the whole process:

> Thirteen thousand years ago, New York was covered by several hundred meters of ice, as it had been for most of the preceding 100,000 years. Then with startling suddenness the glaciers all over Scandinavia and North America disappeared. In Britain the temperature shot up from a summertime value of only 8 degrees Celsius to 18 degrees Celsius, and it did so in a few decades—in a flash, from a historic point of view.[37]

But the temperature quickly began to fall, and not much later than 11,000 years ago:

> The glaciers were back but not yet to their full extent. In northern Britain they covered the mountaintops but did not extend down into the valley bottoms. . . . Then [about] 10,000 years ago there occurred a second warm pulse. Once again within a human lifetime the temperature shot up spectacularly by 10 degrees Celsius, all in a moment from a historic point of view. And this second pulse did the trick. It brought the earth's climate out of the Ice Age of the last 100,000 years into a warm interglacial period which has been essential for the development of history and civilization.[38]

Following the first pulse, "the emergence from cold to warm conditions took only a few decades."[39] And following the second pulse the even more dramatic—indeed, conclusive—warming was accomplished as we have seen, within a human lifetime.

It was therefore natural for Hoyle to investigate what could have caused such sudden and profound changes to global climate:

> My main concern . . . is not so much with the genesis of an Ice Age as with its ending. What, all in a moment, can destroy a situation with a longevity running into tens of thousands of years? Evidently only an exceedingly catastrophic event of some kind, something that would wash out high-level haze, increasing the water-vapor greenhouse sufficiently to send the temperature up almost instantaneously by 10 degrees Celsius. . . . But more still, for unless there was a change from a cold ocean to a warm ocean, the situation would soon return to where it was before. The difference between a warm ocean and a cold one amounts to about a 10-year supply of sunlight. Thus the warm conditions produced by a warm water-vapor greenhouse must be maintained for at least a decade in order to produce the required transformation of the ocean, and this is just about the time for which water, suddenly thrown up into the stratosphere, might be expected to persist there. The needed amount of water is so vast, 100 million million tons, that only one kind of causative event appears possible, the infall of a comet-sized object into a major ocean.[40]

In strong support of Hoyle's reasoning, scientists working completely separately from him have recently reported unambiguous evidence not of one but of *two* major oceanic impacts at around 10,000 years ago—the first in the Tasman Sea, off southeast Australia, and the second in the China Sea near Vietnam.[41] The indications are that these impacts could, between them, have been responsible for the dramatic global warming that took place at that time.

Chandra Wickramasinghe, Hoyle's former student who is now professor of applied mathematics and astronomy at Cardiff University, fully supports the notion of oceanic impacts. In 1998 he told us:

> The natural condition of Earth is one of glaciation, and there's no question about that. . . . Some huge amount of water had to be added in a catastrophic fashion in order to terminate the protracted period of glaciation that existed before 20,000 years ago. . . . I think there's no question that there have been collisions—that Earth's geological record is punctuated by collisions going back to 65 million years ago and earlier.[42]

AGE OF LEO

It is obvious to Hoyle that the impacts that ended the last Ice Age must have been "pretty large, say 10,000 million tons."[43] He admits he was surprised when he first understood that only an episode of this size could explain all the evidence—the surprise coming, as he notes, because there is a habit of mind among scientists to set all such violent events millions of years in the past, and never as recently as 13,000 years ago. In addition, in the 4.5 *billion* years during which we know that Earth has existed, isn't it odd that fragments of a giant comet should have "chosen" to collide with the planet in exactly the period when anatomically modern human beings belonging to the extremely recent species *Homo sapiens*—by then the only surviving species of the genus *Homo*—that is, people exactly like us—should be around to witness it? Then, Hoyle recounts:

> I saw that the answer to this question lies in what is now called the anthropic principle, which says that the fact of our existence can be used to discount all improbabilities necessary for our existence. If history and civilization were caused by the arrival of a periodic giant comet, all accident is removed from our association in time with such a comet. *The arrival of the comet was random but our association with the effects of the comet is not.*[44]

What Hoyle means by the comet "causing" history and civilization is that by ending the Ice Age it created the necessary conditions for human culture and all its achievements to emerge. We, too, see the force of the anthropic principle, but we reach a very different conclusion. In our view civilization does indeed have a dramatic association with impacts from a fragmenting giant comet, but it was not in any way "caused" by those impacts; on the contrary, we suggest that it was nearly destroyed by them. We stick with our scenario of an advanced antediluvian culture that flourished *during* the last Ice Age—in areas of the world that were then hospitable and that are now under as much as 100 meters of water. Our hypothesis is that this great prehistoric kingdom was first massively weakened and then utterly destroyed—leaving only a handful of survivors—by the twin impacts that brought Earth so conclusively out of its long glacial slumber.

As Hoyle and Wickramasinghe have rightly observed, the impacts took place respectively in the eleventh millennium B.C. (between 13,000 and 12,000 years ago) and in the ninth millennium B.C. (between 11,000 and 10,000 years ago). What immediately strikes us about these dates is how closely they coincide with the astronomical Age of Leo, when the constellation of Leo housed the Sun on the spring equinox—generally taken as the period of 2,160 years between 10,970 B.C. (12,970 years ago) and 8810 B.C. (10,810 years ago).[45] As we have seen, this is the "age" that appears to be marked by the lion-bodied, equinoctial Sphinx of Giza—which at the same time draws our attention to Mars through its association with Horus the Red.

The Sphinx has been eroded by *long periods of heavy rainfall* and may actually date back to the eleventh millennium B.C.—as increasing numbers of geologists are now prepared to contemplate.[46] Could its construction have been triggered in some way by the first of those two great cometary bolides that struck Earth in the Age of Leo?

And why should there be a connection with Mars?

Bull of the Sky

FRED Hoyle's evidence of what happened to Earth at the end of the last Ice Age fits Clube's and Napier's theory of a disintegrating giant comet as snugly as the slipper fits Cinderella's foot. To restate the chronology, it is believed that the comet—and there is no known upper size limit for these terrifying objects[1]—settled into an Earth-crossing orbit around 50,000 years ago. For the next 30,000 years it remained relatively intact. Then, about 20,000 years ago, it underwent a massive fragmentation event somewhere along its orbit. From about 17,000 years ago occasional multi-megaton fragments may have collided with Earth—causing some gradual reduction in glaciation—but two especially large and cataclysmic oceanic impacts, one in the eleventh millennium B.C. and one in the ninth millennium B.C., raised global temperatures so much that the Ice Age was brought decisively to an end. These impacts both took place during the astronomical Age of Leo—an epoch, we believe, that is deliberately signaled and symbolized by the Great Sphinx of Giza.

But in its alter ego as Horus the Red, the Sphinx also speaks of Mars, and Mars seems to have its own pyramids and Sphinx—the latter gazing upward from the ravaged and cratered surface of the Red Planet like a veiled human skull.

SIGNAL?

At the end of the previous chapter we asked why there should be a connection between Giza and Mars.

The obvious geometrical and numerical similarities between the "monuments" of Cydonia and the monuments of Giza, and the other strange mythological and cosmological links between the two sites and the two worlds we have reviewed in this book do not, under any circumstances, *prove* a connection.

NASA's fumbling cover-ups, its sustained misinformation campaign, and its generally suspicious behavior concerning the hypothesis of artificial origins at Cydonia do not *prove* there is more going on here than meets the eye.

The work of the AOC (Artificial Origins at Cydonia) researchers has not *proved* that the Cydonia structures are artificial.

Moreover, we ourselves are far from certain—and have remained dubious throughout—about the true provenance of the Martian monuments. They could just be weird geology. They really could. Or they could have been intelligently designed. The only sure way to find out is to *do the science,* and in our view this means nothing less than a manned mission to Cydonia. Improved orbiter photographs are unlikely to settle the controversy—one way or the other—and may just provide more fodder for both the opponents and the supporters of the AOC hypothesis.

Surely the resolution of this matter—on which hinges man's understanding of his place in the cosmos—is too important to be endlessly delayed by such silly shenanigans? Surely it is obvious that if the mathematical data expressed in the Cydonia monuments had turned up in a radio signal from deep space, scientists working on SETI programs would have had a field day (and everyone would have agreed with them), proclaiming they had finally been proved right. Such a clear, coherent extraterrestrial signal would also certainly have been rewarded with a massive investigation involving huge official resources and the concentrated attention of the best scientific minds as humans tried to discover where the aliens were and what they were saying to us. And the investigation would go on even if some skeptics continued to have lingering suspicions that the signal had somehow been generated "naturally" (by freak radio emissions from a star, for example).

We believe that the same sort of response, at the national and international levels, is justified by the Cydonia "signal" even if, on close empirical investigation, it ultimately does prove to be natural. Equipped with radio telescopes and space probes, rapidly evolving technology but a stunted spirituality, our species stands today at what the ancient Egyptian Pyramid Texts call the "portal of the abyss"[2]—literally, on the threshold of the cosmos. If we survive, which is by no means certain, then it is possible that the centuries and millennia ahead will offer us the opportunity of an unparalleled journey of discovery across the galaxy. How can we possibly hope to take advantage of such a fabulous opportunity unless we keep our minds and our imaginations open? How can we possibly learn what the galaxy has to teach if we are not willing to risk disappointment, loss of face, wasted funds, and wild-goose chases?

We therefore repeat that *the science really does need to be done at Cydonia.* It will be expensive, but the funds most certainly can be found. And it is worth doing, irrespective of the final outcome, simply to affirm that we do regard the cosmos with reverent wonder—as our ancestors did—and that we are ready to launch ourselves with curiosity, intelligence, and hope into the deepest mysteries of the galaxy.

But still, why should there be a connection between Giza and Cydonia, between Earth and Mars, and between the comet impacts that ended Earth's last Ice Age with global floods and the massive-impact damage that stripped Mars of half its crust?

We do not *know* that there is any connection at all between the cataclysmic histories of the two planets, and ultimately this is another matter that can only be resolved by empirical tests. We believe, however, that such tests are urgent, necessary, and in the obvious self-interest of humanity, *whether or not* the remains of some sort of lost civilization are to be uncovered at Cydonia. Indeed, they do not even directly concern such a hypothetical and presumably alien civilization—although they may tell us what fate befell it. All that is required is for the first manned landing on Mars to obtain a sufficient variety of rock and dust samples from Martian craters and return them to Earth for analysis. Then radiometric dating and other reliable tests could be carried out that would determine exactly *when* the terminal Mars cataclysm took place.

HYPOTHESIS

As we have indicated several times, we think it is possible that this great disaster, which flayed the planet Mars of its skin, may turn out to have been a much more recent event than scientists have yet imagined. In brief, we propose as a hypothesis for further testing that the giant comet that sprayed the inner solar system with deadly shrapnel around 20,000 years ago did so because it made an extremely close approach to Mars on one of its orbits—closer than Shoemaker-Levy 9 passed to Jupiter in 1994—trespassed the planet's Roche limit and literally exploded into a million pieces.

This would have happened *right on top of Mars,* perhaps at a height of no more than a few thousand kilometers. And the effects, as a vast fusillade of world-killing missiles slammed all at once into the formerly dense atmosphere, the oceans and rivers, the mountains, valleys, and plains of Mars, would have been unspeakably dreadful. Many of these objects, perhaps most of them, would have been larger than 10 kilometers in diameter—each one of them, therefore, packing as much punch as the single fragment of an earlier giant comet that caused Earth's K/T Boundary Event 65 million years ago by making a crater 200 kilometers wide on the edge of the Gulf of Mexico. In addition, since some of the Martian craters exceed 1,000 kilometers in diameter and Hellas has a diameter of 2,000 kilometers, we expect that several of the fragments would have been much larger.

Our theory, therefore, is not so different from the Astra theory outlined in chapter 4. However, Patten and Windsor's work contradicts basic laws of physics when it tries to explain how a former "tenth" planet could have migrated from a stable, circular orbit between Mars and Jupiter into an unstable, elliptical Mars-crossing orbit. Our theory, on the other hand, concerns an object—a periodic giant comet—that one would naturally expect to find in such an orbit, one which has no known upper size limit, which belongs to a class of objects that have been seen to fragment explosively in close proximity to planets, and which has already been implicated in the series of great impacts that ended the last Ice Age on Earth.

In our scenario it was the initial explosion of the giant comet that killed Mars—in a single, phenomenal impact storm. But the rest of the massive swarm of fragments would have missed the Red Planet and continued to travel at high velocity along the comet's original orbit. Since this was a deeply Earth-crossing orbit (with its perihelion close to the Sun and its aphelion beyond Jupiter), we should not be surprised that fragments began to rain down on Earth during the next several thousand years—not killing it, as they had Mars, but nevertheless causing profound and dramatic changes.

A SPECULATION

It is permissible, sometimes, to speculate and we offer the following as no more than an *amuse gueule,* a harmless speculation, intended to entertain. It is a kind of artifact of our imaginations that arises every time we look anew at the image of the Face on Mars and at the geometrical structures that seem to have been arranged so purposefully around it on the Cydonian plain.

The math feels like a message.

The peculiar interlinkages to Giza and to Teotihuacan don't feel accidental.

The latitude games played at all three sites do feel as though they share the same designer.

Last but not least, some of the structures of Cydonia stand immediately beside and even inside impact features—including, for example, an intact pyramid, unencumbered by ejecta and not at all damaged, poised on the very edge of a crater rim.[3] Such anomalies suggest to us that the monuments must have been built *after* the terminal Mars cataclysm, not before it.

Our hunch, therefore, is that Cydonia is indeed some sort of signal—not a radio broadcast intended for the entire universe, but a specific directional beacon transmitting a message that is intended *exclusively for humankind.*

To receive that message we have to prequalify.

We have to be able to look at Mars closely, which means high technology. But we also have to have the intelligence and open-mindedness, the vision and the spiritual humility to accept that even a dead planet can speak to us.

In short, humanity has to be able to see Cydonia, to realize what it is and to act on what it says.

Who might have devised such a message? And how could they have arranged to express it in a distinctive "architectural/geometrical code" that would much later turn up on Earth in the pyramids and the Great Sphinx of Giza and other terrestrial sites such as Stonehenge and Teotihuacan?

Could it possibly be that the builders of Cydonia also contrived to exert an influence upon the early civilizations of Earth? Were they somehow involved here, perhaps during the darkest midnight of prehistory, perhaps even long before the biblical flood? Could this explain why there seems to be a lingering and tantalizing "memory" of Cydonia imprinted upon the design plan of the Giza complex and why not only the Sphinx but also the Arab city of Cairo that grew up around it were called by names meaning "Mars"?

Lastly, what about the content of the message of Cydonia?

We go on instinct, nothing more, but in our speculation it is a warning that a Mars-like doom lies in wait for the Earth unless we take steps to avert it—a doom that could spell the end not just of individual species, not just of human civilization, but of all human beings and of all life on this planet. That is why the message is addressed exclusively to us—because we are its potential beneficiaries. That is why it is expressed in a language of architecture, geometry, and symbolism that strikes a chord with humans. And that is why there is indeed a deep and ancient connection between Earth and Mars, anchored to certain astronomical monuments that were designed, from the very beginning, to awaken us at the eleventh hour.

A PATTERN OF IMPACTS

Let us now return to the giant comet and recall its life cycle after it descended from the galaxy into the inner solar system:

- 20,000 years ago: explosive fragmentation beside Mars
- 13,000 to 12,000 years ago: major bombardment of Earth; glaciers retreat
- 11,000 to 10,000 years ago: second major bombardment of Earth; Ice Age ends

None of the astronomers who opened up this extraordinary field of study in the past twenty years are under any illusions that the menace to Earth ended with the Ice Age cataclysms. On the contrary, they are certain that fragments of the giant comet have continued to fall among us.

The detailed investigation into the matter by Fred Hoyle and Chandra Wickramasinghe has yielded information from temperature records and other sources suggesting that major impacts—though none as severe as those that occurred during the Age of Leo—have continued at sporadic intervals throughout human history. According to these two scientists, the evidence suggests that there were episodes of chaos, disruption, and rapid climate change at around 7000 B.C., 5000 B.C., 4000 B.C., 2500 B.C., 1000 B.C., and A.D. 500—in each case lasting for several decades or even a century and involving repeated collisions with multiple fragments of at least Tunguska size, *up to a rate of 100 per year.*[4]

Duncan Steel believes the rate of impact may at times have been much higher and calculates that in such episodes

> Cataclysms visit wide areas of the planet due to the coherent arrival of many impactors in a few days. *It is entirely feasible that within those few days the earth could receive hundreds of blows like that of the Tunguska object.* [author's emphasis][5]

THIRD MILLENNIUM B.C.

Post–Ice Age history has also been looked into by other researchers, who agree that many anomalies are explained by the notion of an irregular rain of fragments repeatedly disrupting cultures all around the world.

The second half of the third millennium B.C., for example, from 2500 B.C. to 2000 B.C., appears to have been a turbulent and dangerous period during which surprising numbers of formerly well-established civilizations inexplicably collapsed or underwent a period of chaos and disintegration. After studying more than five hundred excavation reports and climatological studies, Dr. Benny Peiser of Liverpool John Moore's University has demonstrated that all of the affected civilizations "suffered huge changes in climate at exactly the same time."[6] These disasters occurred "in the Aegean, Anatolia, the Near and Middle East, Egypt and North Africa, and large parts of Asia."[7] There was also a related catastrophe as far afield as eastern China.[8]

The Indus Valley civilization in the northwest of the Indian subcontinent was one of the victims, vanishing mysteriously.

Egyptian civilization survived the climatological upheaval but preserved memories of intense heat, violent floods, and the abrupt desertification of previously lush agricultural lands.[9]

In the same epoch the Akkad empire of Mesopotamia and Syria collapsed amid floods and evidence of a major cataclysm—hitherto presumed to have been a large earthquake—which was confirmed by researchers in 1997 to have been an impact.[10] Marie Agnes-Courty of the French Center for Scientific Research found microspherules of a calcite material—unknown on Earth but abundant in meteorites—scattered across an area of thousands of square miles in northern Syria in soil samples and archaeological deposits dated to 2350 B.C.[11] She also uncovered evidence of gigantic regional fires in the form of a thick deposit of black carbon.[12]

Parallel research has identified at least seven other impact craters around the world "which were formed within a century of 2350 B.C."[13] And Professor Mike Baillie, a paleoecologist at Queen's University, Belfast, reports that his studies of tree rings have uncovered evidence of widespread ecological catastrophes at this date.[14]

THE TAURID MYSTERY

In the second half of the third millennium B.C., while all these events were unfolding, astronomical calculations show that the orbit of Earth would have intersected the core debris of the particularly massive and widely diffused Taurid meteor stream—so called because it produces showers of "shooting stars" that look to observers on the ground as though they originate in the constellation Taurus.[15] The stream sprawls completely across the Earth's orbit—a distance of more than 300 million kilometers—cutting it in two places so that the planet must pass through it twice a year: from 24 June to 6 July and again from 3 November to 15 November.[16] Since Earth travels more than 2.5 million kilometers along its orbital path every day, and since each passage takes approximately 12 days, it is obvious that the Taurid stream is *at least 30 million kilometers wide, or thick.* Indeed, what Earth encounters during these two periods is best envisaged as a sort of tube or pipe of fragmented debris.

Even though it is one of the most intense of all the annual meteor showers,[17] the encounter from 24 June to 6 July (which peaks on 30 June) cannot normally be seen with the naked eye—only with radar and infrared equipment—because it occurs during daylight hours. But the encounter from 3 November to 15 November is visible at night. The *Collins Guide to*

Stars and Planets tells amateur astronomers where to look in the constellation of Taurus:

> The meteors radiate from a point near epsilon Tauri, reaching a maximum of about 12 meteors per hour on 3 November.[18]

The reader will recall from chapter 23 that in the ancient Egyptian sky-ground plan the two pyramids of Dashur, supposedly built at around 2500 B.C., correlate with the positions of two stars in Taurus—the Red pyramid with Aldebaran and the Bent pyramid with epsilon Tauri. We note that the date of 2500 B.C. was toward the end of the astronomical Age of Taurus (when the Sun on the spring equinox rose in the constellation of Taurus, roughly from 4490 B.C. to 2330 B.C.). We have seen that the Sphinx serves as an astronomical marker for the Age of Leo (10,970 B.C. to 8810 B.C.)—the epoch that experienced the gigantic impacts that ended the last Ice Age. We have seen that Earth appears to have been shaken by another series of bombardments during the period 2500 to 2000 B.C.—the epoch of pyramid construction in Egypt. And we saw in chapter 17 that the Benben stone, the sacred cult object of the Heliopolitan priests who served the pyramids, was almost certainly an "oriented" iron meteorite.

Could there be a connection among (a) the bombardments and the Taurid meteor stream and (b) observations of Taurid meteors at around 2500 B.C.—which must have been spectacular as Earth neared the core of the stream—and (c) the construction of the pyramids of Egypt?

STONEHENGE

We have no doubt that the pyramids—and other ancient megalithic structures all around the world—were religious and spiritual buildings; nevertheless we do not object to the notion that they might also have had a number of more practical, or even scientific uses. The ancients did not make the distinction between science and spirit that we do today, and we suspect that the Heliopolitan cult required its initiates to cultivate what can only be described as a scientific knowledge of the sky. We therefore see no contradiction at all between the practical observational and mathematical functions of a monument and its overriding spiritual and transformational purpose.

Nor are we the first to suggest that among the complex motives in the long-term development of certain mysterious ancient sites there may have been a special interest in meteor showers.

Dr. Duncan Steel is the director of Spaceguard Australia.[19] We have referred to his work and discoveries frequently in these pages. It is his theory that the primary axis of Stonehenge in England, which lies 33 degrees of longitude west of Giza, was not originally designed to target the summer solstice sunrise (the most widely accepted view) but was targeted instead on the rising of the Taurid meteor stream.[20] This was done during the "preliminary" period, which archaeologists refer to as Stonehenge I—roughly from 3600 B.C. to 3100 B.C.—and the great megaliths that we see today were laid out to conform with the same axis. The period of megalith construction is well dated at 2600 to 2300 B.C., when the bluestones and the sarsens (the famous "goalposts") were erected[21]—a period that overlaps curiously with the pyramid age in Egypt and with the worldwide episode of bombardment in the second half of the third millennium B.C. But such bombardments are by their nature recurrent—at unpredictable intervals—and can be sustained over centuries on each occasion. Steel has produced evidence that an earlier episode occurred at the time of Stonehenge I, in the second half of the fourth millennium B.C.[22]

Steel's case, which is solidly based on dynamical studies and backtracking of trajectories within the Taurid stream, is that the disintegrating giant comet that has shadowed Earth like a vampire or a ghoul for the past 20,000 years underwent one of its spectacular fragmentations some time in the fourth millennium B.C. This was when the Taurid meteor stream was spawned and sent swarming through space on its Earth-crossing orbit—a swarm, as we shall see, that consists not only of meteorites and dust but that also incorporates an inert, near-invisible mass of asteroids and several active comets. One of these, periodic comet Encke, still well-known to modern astronomers, was highly violatile and would have been spectacularly visible with a fully developed "coma" and tail by about 3600 B.C. At the same time, as other fragments worked their way down to Earth, humans would have witnessed "intense meteor storms" and would almost certainly have been subjected to sustained periods of heavy bombardment by massive lumps of debris resulting in "multiple Tunguska-type events."[23]

In a nutshell, what Steel is claiming is that the Stonehenge axis, with its distinctive northeast orientation (he believes only coincidentally close to the rising point of the sun on the summer solstice) was laid out as a kind of "early warning system for cosmic impacts"[24]:

> From Stonehenge I . . . as the comet neared the earth it would have appeared to rise in the evening with a huge bright stripe [the Taurid meteor trail] crossing much of the sky, originating in the northeast. Passage through the trail would then have resulted in celestial fireworks (and maybe worse); afterward the comet and trail would have passed in the direction of the Sun, partially blocking sunlight for a few days. . . . It is suggested that Stonehenge was built . . . to allow the prediction of such events.[25]

ENCKE

Shooting stars are harmless—nothing more than tiny meteors burning up in the atmosphere—so why should anyone be afraid of a meteor trail?

In the case of the fifty or so distinct and separate meteor streams that have now been discovered by astronomers—the Leonids, the Perseids, the Andromedids, etc.—the answer to this question is that in most cases there is probably no danger and nothing to fear.[26] As most of the particles they contain are indeed tiny, they represent no threat to Earth.

But it is quite a different matter with the Taurids. As Steel, Asher, Clube, Napier, and their colleagues have demonstrated, the reason is that the Taurid stream is filled to overflowing with other much more massive material, sometimes visible, sometimes shrouded in clouds of dust, and all of it flying through space at tremendous velocities and intersecting Earth's orbit, regular as clockwork, from 24 June to 6 July and again from 3 to 15 November. Year in, year out, for a period of more than 5,000 years, comet Encke and all the other contents of the stream were spawned from the continuing disintegration of the vastly larger interstellar giant.

The gradual revelation of the truly dark and horrendous character of the Taurid stream results from the work of astronomers going back over half a century—work that members of the public remain largely unaware of, although it raises question marks over the future of civilization.

The fundamental discovery was made in the 1940s when the American astronomer Fred Whipple was the first to point out the intimate relationship between the Taurid stream and the comet Encke, which lies at the heart of Steel's Stonehenge theory. It has a highly elliptical Earth-crossing orbit of just 3.3 years—a shorter orbit than any other known periodic comet[27]:

> Encke is about five kilometers across. . . . It may, therefore, be correct to think of it as the parent of the stream. On the other hand, there may well be one or more dormant comets in the stream that we have yet to identify and that may exceed Encke in size.[28]

By 1998, as we shall see in the next chapter, increasingly sophisticated astronomical surveys involving radar and the radio telescopes at Jodrell Bank, the Spacewatch telescope at Kitt Peak in Arizona, and the highly successful Infrared Astronomical Satellite had begun to reveal the full extent of the problem.

26

Dark Star

IF THE overall climate of our globe should once again improve," warn Victor Clube and Bill Napier, "as it is doing during this century, and has done every few centuries since the end of the last Ice Age, there may be only the dimmest perception of an approaching nadir. We may be unaware that the cosmos is simply delaying the next input of dusty debris, alarm, destruction, and death. A great illusion of cosmic security thus envelops mankind, one that the 'establishment' of church, state, and academe do nothing to disturb. Persistance in such an illusion will do nothing to alleviate the dark age when it arrives. But it is easily shattered: one simply has to look at the sky."[1]

After everything that we have learned while writing *The Mars Mystery* we are frankly baffled that organizations like NASA that receive public funds to "look at the sky" are using so little of that money to investigate the dangers of serious collisions with objects on Earth-crossing orbits. While disposing of a budget of $13.8 *billion* annually, NASA spent less than a million dollars during 1997 supporting near-Earth asteroid and comet surveys.[2] Britain in the same year spent just £6000—about $10,000—making it clear when it did so that this was a one-shot grant that was unlikely to be repeated.[3]

"Such a singularly myopic stance," comment Clube and Napier, "may place the human species a little higher than the ostrich, awaiting the fate of the dinosaur."[4]

Or, as Sir Fred Hoyle sees it:

> It could be seen as curious that society would seek to investigate distant galaxies while at the same time ignoring all possibility of serious impacts with Earth, surely a clear example of amnesia in action.[5]

The minimum response, says Hoyle, and only a first step, would be "to compile a catalogue of all objects of appreciable size in Earth-crossing orbits. For this a space telescope is needed. But not as large or expensive as the Hubble telescope. One with an aperture of a meter should be adequate, at any rate initially."[6]

Even this modest demand, set out in 1993 by an eminent astronomer, had not been met by 1998—when there was still no dedicated space telescope looking for near-Earth objects. Yet the utility of such a satellite for detecting potentially dangerous comets or asteroids that terrestrial observers would be unable to see—perhaps until it was far too late to mount any effective response—has been obvious since the launch of the Infrared Astronomical Satellite (IRAS) on 27 January 1983. A cooperative venture involving public funds from the United States, the Netherlands, and the U.K., its primary objective was to carry out a deep-space survey that ultimately produced a catalogue of a quarter of a million infra-red sources "including stars, galaxies, dense interstellar dust clouds, and some unidentified objects."[7] But during its ten months in orbit (the mission ended on 23 November 1983 when the satellite's supply of coolant ran out), IRAS also spent a little time looking at near-Earth space. There it discovered five new comets undetected by terrestrial astronomers (comets are very hard to see when they approach Earth from the direction of the Sun). One of these, IRAS-Araki-Alcock, was observed by the satellite in May 1983. The reader will recall that it passed within 5 million kilometers of Earth—the closest-known approach by any comet since the visitation of comet Lexell in the eighteenth century.[8]

What else might IRAS have seen swarming around Earth if it had turned its camera on the comet threat full-time? Or if it had been designed and equipped to observe for longer than ten months?

As rational people who have looked at the evidence with open minds, we genuinely cannot understand why NASA, the organization that is best placed and best funded to do something about the impact threat, has so far done so laughably little. It reminds us of the way that the same organization has responded to the extraordinary challenge of the "monuments" of Mars. In both cases there is a mass of intriguing evidence—whatever it may ultimately prove to mean. And in both cases NASA has steadfastly minimized it.

Is there some sort of conspiracy going on to keep the truth from us about the terminal Mars cataclysm and how it concerns Earth?

On balance we prefer to think *not.*

What we see is a mind-set, here, not a conspiracy.

And yet . . .

To be perfectly honest, we will always have a lingering suspicion that there could be something dark and dreadful going on behind the scenes, something much bigger, and much more awful, than a mere conspiracy. The universe is mysterious. Reality itself is mysterious. No human has any true idea whether life has any transcendant purpose or not, whether there is life after death, whether there are such entities as absolute good and absolute evil.

We therefore see no reason to reject out of hand the teaching of the ancients on these matters—which is that man is the fulcrum of a great cosmic conflict. Opposing forces of darkness and light, nihilism and celebration, and hate and love struggle to win victory over man's soul, because such a victory will decide the fate of this created universe and define the character of all universes yet to be formed. The light gains the upper hand when reason and mind are cultivated among humans, allowing them to turn their attention away from purely material concerns and cultivate the spirit. The darkness responds by interfering in the world to destroy mind and reason and thus frustrate humanity's spiritual promise and ultimate role in a wider redemption. Again and again, the ancients said, when former races of men had risen to a high level they were cruelly punished and forced to return to a low state.

Thus the Gnostic texts, written down in Egypt in the early centuries of the first millennium after Christ, tell us that the global cataclysm remembered as the Flood of Noah was not inflicted by "God" to punish evil—as the Bible claims—but was worked by the forces of darkness to punish antediluvian humanity for having aspired to a high state of scientific and spiritual development and "to take the light" that was growing among men.[9] This the darkness in very large part succeeded in doing. Although there were survivors, most were thrown

> into great distraction and into a life of toil, so that mankind might be occupied by worldly affairs, and might not have the opportunity of being devoted to the holy spirit.[10]

Plato's tale of lost Atlantis likewise laments that whenever civilization reaches a high level, opening the way for study and contemplation and matters of the spirit, "the periodic scourge of the deluge descends, and spares none but the unlettered and uncultured" so that human beings forget the past, and all that they have learned, and must "begin again like children."[11]

Plato's narrative rather curiously links the deluge to a "thunderbolt" and to "a variation in the course of heavenly bodies and a consequent widespread destruction by fire of things on Earth."[12]

So with global floods, followed by fires and a remembered connection to thunderbolts and the heavens, what we have here sounds like the effects of a multiple-impact bombardment with white-hot bolides falling from the sky and bursting in the air, and others plowing into distant oceans and creating vast tsunamis capable of tearing across continents—sparing, as Plato puts it, only "the herdsmen and shepherds in the mountains."[13]

After looking at the cratered and devastated hulk of Mars, there can be no doubt in anybody's mind that this planet was destroyed by a scourge from heaven. All its potential, whatever it might have become, whatever life or civilization or miracles it might have been home to, stopped right there, right then, and it was all over.

The universe is infinitely mysterious, infinitely various. We therefore do not find it impossible to imagine how some monstrous cosmic intelligence that feeds on negativity and darkness might be nourished and fattened by such an unspeakable tragedy. Indeed it is a supernatural force of exactly this kind that is envisaged in the Gnostic texts as unleashing the flood upon mankind in order to deprive us of our "light."

How much deeper the universal darkness would become if that little light could be snuffed out forever.

Yet if the Gnostics were right the darkness *cannot* triumph on its own. It needs and seeks our help, our willingness—our complicity—to achieve the destruction of the light.

ORBITING IN THE TORUS

Prolonged studies of the Taurid meteor stream by dedicated astronomers working on their own time at many different observatories—and borrowing time on telescopes dedicated to other purposes—have begun to produce a picture of a threat that could indeed bring down the darkness. Cloaked in billions of tons of swirling dust, and surrounded by dozens of kilometer-size asteroids, it appears that a huge, inert, almost invisible comet may lie at the core of the stream—a larger fragment from the explosion that spawned Encke more than 5,000 years ago.[14]

In the last chapter we compared the Taurid stream to a pipe or a tube of rushing debris laid across the path of Earth. But since the stream in fact extends all the way around comet Encke's elliptical orbit (with all of its contents in continuous rapid motion along that orbit) its true form is that of a tube formed into an ellipse. The shape, in other words, is a three-dimensional ring like a doughnut or a quoit, but with a cross-section of 30 million kilometers. The correct term for such a shape is a *torus.*[15]

What else is orbiting in the torus along with "shooting stars" and the 5-kilometer nucleus of periodic comet Encke?

Thirteen Earth-crossing Apollo asteroids, all more than one kilometer in diameter, have been firmly identified.[16] Based on calculations widely accepted among astronomers concerning the ratio of discovered to undiscovered asteroids sharing the same orbit, Clube and Napier conclude from this data that there must be a total of

> between one and two hundred asteroids of more than a kilometer diameter orbiting within the Taurid meteor stream. It seems clear that we are looking at the debris from the breakup of an extremely large object. The disintegration, or sequence of disintegrations, must have taken place within the past twenty or thirty thousand years, as otherwise the asteroids would have spread around the inner planetary system and be no longer recognizable as a stream.[17]

In addition to comet Encke, there are at least two other comets in the stream—Rudnicki, also thought to be about 5 kilometers in diameter, and the mysterious Apollo object named Oljato, referred to in chapter 21, which has a diameter of about 1.5 kilometers.[18] Initially believed to be an asteroid, this extremely dark Earth-crossing projectile has recently begun to show signs, visible in the telescope, of volatility and outgassing, and most astronomers now regard it as an inert comet that is in the process of waking up.[19] Comet Encke itself is known to have been inert for a long period, until it suddenly flared into life and was seen by astronomers in 1786.[20] It is now understood to alternate regularly, in extended cycles, between its inert and volatile states.

Clube and Napier have backtracked the orbits of Encke and Oljato and found that they were nearly identical until about 10,000 years ago[21]—roughly the epoch of the second great Ice Age impact. Since we know that Encke was itself the product of a fragmentation event over 5,000 years ago[22]—at which time it separated from a larger and as yet unidentified parent object—the likely conclusion is that Oljato was also a fragment of that original parent object, which had separated as a result of an earlier disintegration:

> It is possible there was a major disintegration of the prime body then, with much debris created of which Comet Encke and Oljato are the largest known bodies, followed by similar disintegrations of the other comets and asteroids of the stream.[23]

There is what the astronomers call a great deal of "fine structure" within the Taurid stream as a whole—that is, distinct groups of objects can be identified orbiting within the 30-million-kilometer wide tube of the torus. Backtracking these orbits, Clube and Napier note that the meteor group called the northern Taurids seems to have broken away from comet Encke, or perhaps a Taurid asteroid, about a thousand years ago. They conclude that the whole complex, meaning the assorted contents of the entire torus,

> seems to be undergoing avalanching self-destruction as the debris accumulate and collide. . . . This unique complex of debris is undoubtedly the greatest collision hazard facing Earth at the present time. It is likely that *hundreds of thousands of bodies, each capable of yielding a multimegaton explosion on Earth, are orbiting within the stream.* [author's emphasis][24]

MULTIPLE STREAMS

It is well understood by astronomers that the largest and densest bodies within any stream will concentrate toward its center,[25] and it has also been established that the Taurid stream does have a dense core, along the edge of which orbits comet Encke[26]—towing in its wake a thick disjointed "trail" (as distinct from tail) of debris first observed in 1983 by the invaluable IRAS satellite.[27] It is also obvious that the farther one travels from the core the more diffuse, small, and harmless the orbiting particles are likely to be.

In the case of the Taurids this picture is complicated by the fact that two other massive streams of material, again arrayed in the form of gigantic elliptical tubes, are flung out in orbits parallel to the central torus, one stream closer to the Sun at perihelion and one farther away. These are jointly called the Stohl stream (after their Czechoslovakian discoverer) and are believed to have been created by further spectacular disintegration, probably at around 2700 B.C., of a large fragment of the parent giant comet.[28] Clube and Napier calculate the mass of meteorites within the Stohl stream as "10 or 20 million million times a million grams" and estimate that "the mass of co-orbiting asteroids is likely to be the same." Adding in gas and dust that have been lost with the passage of time, they conclude that the mass of material is roughly equivalent to that of a body of 100 kilometers in diameter.[29]

Further complicating the picture is a completely separate though narrower torus that has the same dynamical characteristics as the orbits of the Taurid and Stohl streams, and which must also once have been part of the same very large progenitor that spawned Encke. What has happened, however, as a result of some powerful event at an unknown date thousands of years in the past, is that the *plane* of its orbit has been rotated through approximately 90 degrees to the main Taurid and Stohl streams.[30] This is the so-called Hephaistos group and includes the Apollo asteroid Hephaistos after which it is named—as the reader will recall, Hephaistos has a diameter of 10 kilometers,[31] as big as the K/T impactor that killed the dinosaurs 65 million years ago. Five other kilometer-plus asteroids have also been observed traveling with Hephaistos as well as the usual mountains of dust and loosely graded debris.[32]

The implication is that future discoveries are likely to yield at least another fifty asteroids of kilometer size spread out along the Hephaistos orbit.[33]

THE UNDETECTED COMPANION

So the overall picture of the Taurid hazard can now be seen to include four separate but intimately related streams of material—the two Stohl streams, the Hephaistos group, and the main Taurid stream with comet Encke as the most visible object. All of these streams derive from the fragmentation of the same original giant comet, and all are on "inter-nested" near-Earth orbits so disposed that our planet passes from one to another during the year—and indeed spends a total of more than four months of each year actually immersed in them.[34]

Each crossing must be hazardous: we already know that there are very large and menacing objects rushing along in these streams, and it is obvious that many more remain to be detected. It is the Taurid stream itself, however, that Clube and Napier ultimately highlight as the deadliest collision hazard faced by Earth.

This is because their research, now supported by a growing number of astronomers and mathematicians, has highlighted the most terrible danger of all—in the form of an *undetected companion* to comet Encke that is believed to be orbiting in the very heart of the stream.[35] The suspicion that such an object could exist goes back as far as 1940, when Fred Whipple showed that several groups of meteor orbits could not be explained in any other way other than as an ejection of debris from an exceptionally large object in an inclined orbit close to that of comet Encke.[36]

Further evidence accumulated since Whipple's time has led the researchers to conclude that such an object does indeed exist. They believe that like Encke and Oljato, the undetected companion is a comet that sometimes—for very long periods—is able to shut itself down.

This happens when pitch-like tars that seethe up continuously from its interior during episodes of outgassing become so copious that they coat the entire outer surface of the nucleus in a thick, hard shell and seal it off completely—perhaps for millennia.[37] On the outside all falls silent after the incandescent coma and tail have faded away and the seemingly inert object tears silently through space at a speed of tens of kilometers per second. But at the center the nucleus activity continues, gradually building up pressure. Like an overheated boiler with no release valve the comet eventually explodes from within, breaking into fragments that can become individual comets or crash into planets.

We have seen in chapter 22 that the nucleus of Halley's comet is so black that it reflects only 4 percent of incidental sunlight.[38] It is suspected that in its inert state the nucleus of Encke's undetected companion may be even blacker—perhaps among the blackest objects in the solar system. Since it would also be surrounded by a dense cloud of meteoritic dust it is legitimate to think of it as a sort of cosmic Stealth missile.

It is difficult to estimate the exact size of this frightful Earth-crossing companion or what its future orbital parameters might be. Nor can we be certain how many other large fragments could be swirling along with it, also cloaked in meteoritic dust. Despite these uncertainties, some attempts have been made, and in 1997 the Italian mathematician Emilio Spedicato reported certain grave conclusions.

The object, he calculated, might be *30 kilometers in diameter.*[39]

Moreover:

> Tentative orbital parameters which could lead to its observation are estimated. It is predicted that in the near future (around the year 2030) Earth will cross again that part of the torus which contains the fragments.[40]

SHIFTING ORBITS

We should hope, devoutly, that Spedicato is not correct about the date—because a collision with an object 30 kilometers in diameter will certainly end all human life, and may indeed unleash sufficient impact energy to sterilize the planet of all life. Some of the astronomers who have built up the evidence we have on comets feel reassured that the fateful intersection will probably not occur for another thousand years.[41] Victor Clube is one of them. Others, notably Fred Hoyle and Chandra Wickramasinghe, have indicated that according to their calculations another episode of bombardment *is* on the way and can be expected to hit during the coming century.[42]

The problem is that nobody can really be sure. Earth's orbit is constantly, though minutely, shifting in shape, becoming now more or less eccentric (elliptical), now more or less circular. At the same time its perihelion and its aphelion gradually "precess" around the orbit—that is, move backward in relation to the direction of principal rotation. Meanwhile the same celestial mechanics are at work on the torus. The effect is that the points of intersection of the two orbits vary considerably from epoch to epoch, and not only that but also the area of the torus through which Earth passes.[43] A transit through the edge of the stream is likely to be tranquil, with consequences limited to nothing more than shooting stars. On the other hand, a transit through or close to the core could result in an almost unimaginable disaster—especially if there should be a collision between Earth and Encke's dark companion.[44]

Where are we now?

CLUES OF JUNE

Once again, astronomers have different views. Nevertheless, they all point to one curious thing—a pattern involving the month of June.

We have seen that the Taurid shower produces visible meteors when Earth passes through it from 3 November to 15 November each year, but produces a much larger and more virulent storm of debris, invisible to the naked eye, between 24 June and 6 July—with a peak on 30 June. Because of the relative positions of Earth and the Sun, this is a period in which large projectiles could theoretically creep up on Earth with the Sun behind them and fall upon us before anybody had really noticed.

On 25 June 1178, it was exactly such a projectile, an Apollo asteroid or a comet fragment 2 kilometers in diameter, that hit the Moon, creating the gigantic Giordano Bruno crater (see chapter 18). It was extremely fortunate, and indeed a miracle, that it did not hit Earth—as Earth is in the same area of space as the Moon and makes a much bigger target.

In chapter 18 we also presented two other essential clues:

- On 30 June 1908, a much smaller fragment of the disintegrating comet exploded in the air above Tunguska, flattening 2,000 square kilometers of forest and causing huge earth tremors hundreds of kilometers away.
- From 22 through 26 June 1975 the Moon was splattered with a sustained barrage of ton-sized boulders.

Astronomers are now generally agreed that all of these impacts were related to passages of Encke's comet, which travels particularly closely to the June/July Taurids, and were caused either by subdividing fragments falling off it or by other objects orbiting close to it that were cast down into the Earth-Moon system.[45] Since we know that Encke orbits near to the core, and thus to "the unseen companion," it is evident these past encounters could have been far worse.

And what of future encounters?

The vision that haunts us is of that dark, dark nucleus, wrapped in its veil of dust, throwing before it a swarm of asteroids.

As Clube and Napier warned as far back as 1990 (apparently to no avail, as there has been no change in public policy):

> An asteroid in a Taurid orbit, carrying 100,000 megatons of impact energy, coming out of the night sky [during the November crossing of the stream] would be visible in binoculars for about six hours before impact. By the time it was a naked-eye object it would be at most half an hour from collision. In its final plunge it would be seen as a brilliant moving object for perhaps 30 seconds. One needs more time than this to prepare for winter.[46]

If such an asteroid came in broad daylight during Earth's encounter with the Taurid stream in late June—the time at which a collision is also most likely to occur with comet Encke or its dark companion—then it would not be seen at all unless there were a satellite in the sky equipped with an infrared camera.

HELL-WORLD OF OUR OWN MAKING

Humanity today faces two strange and powerful firsts:

- The first time, in the history we remember, that a disaster looms with the potential not just to destroy some of mankind but to destroy all of mankind—all human promise, all human potential, forever.
- The first time, again in remembered history, that our species has the science and the technology to avert such a disaster—if it has the will.

We have received unambiguous warning signals—from the fate of Mars, from our growing understanding of the effects of impact-cratering on Earth, from the pattern of known large-body Taurid impacts on the Earth-Moon system during the second millennium A.D., and from the apocalyptic crash of comet Shoemaker-Levy 9 in 1994.

Reason and intuition concur. There is real danger here.

Yet next to nothing is being done about the danger and Clube's warnings, Sir Fred Hoyle's warnings, and the warnings of all the other eminent men and women who have seen the threat. They are ignored.

We suspect that the first half of the third millennium will be a defining epoch in the story of mankind that will require not just a change of policy, or a change of strategy, or a change of budgetary priorities—though it will certainly require all those things—but above all else *a change of heart.*

To a great extent, the ancients said, we define our own reality through the choices we make. Yet what we have made with those choices at the end of the twentieth century is well on its way to becoming a hell-world.

What has happened to the human soul when a man, saying that he is acting in the name of God, is so in love with hatred that he can smash out a baby's brains against a wall and cut the throat of that child's mother? Such events had become routine in Algeria at the end of the second millennium.

What has happened to the human soul when adults—men and women—are so in love with evil that they gain sexual pleasure from the kidnap, torture, rape, and murder of children? Such horrors had become routine in Europe and the United States at the end of the second millennium.

What has happened to the human soul when a man is so in love with his own ego that he can dash concentrated sulphuric acid into the face of a teenage girl—eating away her flesh, blinding her and burning her skin forever to a crisp—simply because she has refused to marry him? At the end of the second millennium such acts of focused malice and wickedness had become routine in Bangladesh, inflicting lifelong shame, misery, and suffering on hundreds of girls every year.

We will not continue with lists of individual and mass atrocities that could extend to hundreds of volumes—as everybody knows. We simply wish to suggest that a species that is so drawn toward the darkness is unlikely to be able to meet the challenge of the galaxy. Indeed, we seem to have proved we cannot meet it during the first decades of our discovery of Mars and by our failure to take any interest in the protection of our own precious and irreplaceable planet, which—so far as we know—may be the last remaining home for life in the universe.

THE ARROW AND THE CHOICE

To deal with the impact threat effectively would require a grand international project, with limitless resources and limitless good will, bringing together the best minds in the world and asking them to consider nothing else but the safety of the planet and the salvation of their fellow humans. Deflecting asteroids and fragments of dormant comet that could be up to 30 kilometers in diameter would be a high-precision task, since it is obvious that any error might make the trajectory of the incoming object *more* rather than less dangerous. Probably it is at the very edge, or just beyond the edge, of what our science today is capable of achieving. It sounds impossible. And yet, if you stop to think about it, something of the kind is already being done to achieve far less worthwhile objectives. The world's armed forces, for example, are a kind of "grand international project," with limitless resources, bringing together the best minds of all nations and asking them to consider nothing else but ways to spread mayhem and misery, to bombard and to poison, and to inflict death and destruction on their fellow humans.

So really what is involved here are the kinds of choices that societies make about what they want to do with their resources, not a problem over the resources themselves. Yet we can hardly imagine any society in the world as it is today, let alone one of the major powers, actually deciding to switch significant funds from defense and aggression against humans to the defense of the planet.

This is why we are sure that what will ultimately be required, if there is time and if the threat of cosmic impacts is to be overcome, is that human beings should reinvent themselves in the twenty-first century—reinvent themselves entirely. We even wonder whether a grand project to save the earth might not in itself act as the necessary catalyst for such a change. Indeed, in its own way, with almost no official trappings, we have seen that the project has already begun—depending on the energy and initiative of a loose network of astronomers and other scientists volunteering their time in many different countries for the good of humanity.

As an old saying, attributed to Hermes, has it:

> Death is like an arrow that is already in flight, and your life lasts only until it reaches you.[47]

What the astronomers have shown us is an arrow in the sky, aimed at Earth, that has been flying toward us for five million years.

Yet this arrow need never arrive. Life and light and laughter and the quest for sacred knowledge need not be rubbed out. The darkness need not be fed with more suffering and nihilism. Magic and mystery can be renewed. And the wasteland can be healed.

We are defined by our choices.

And this choice is ours.

Appendix

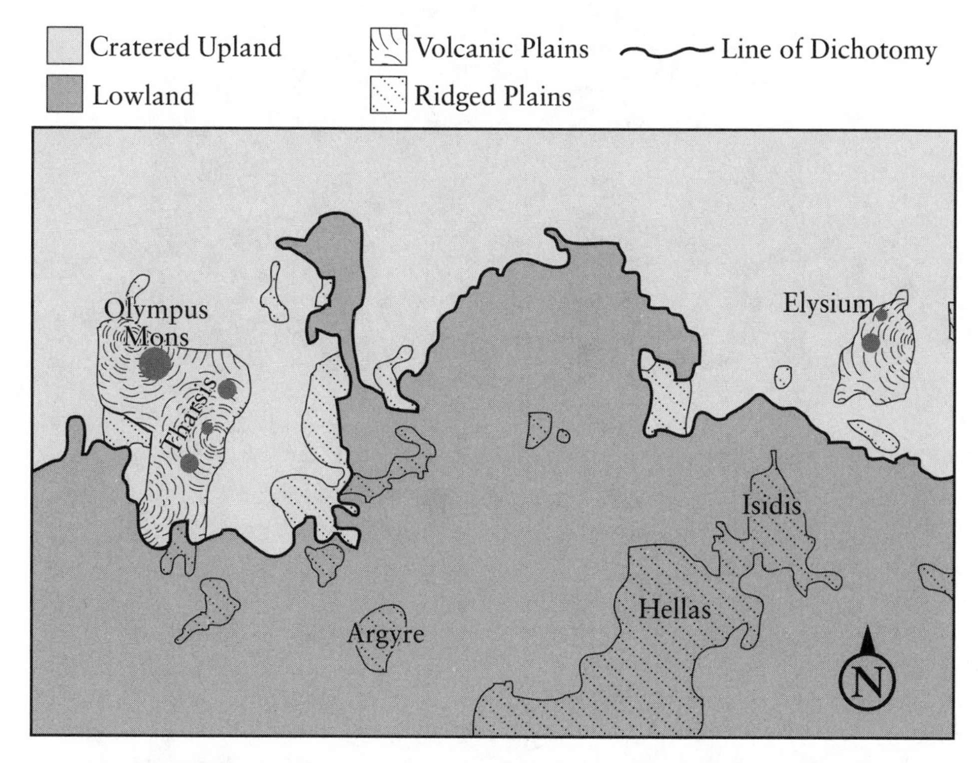

The "line of dichotomy" is a dramatic cutoff point that separates the heavily cratered southern highlands from the more sparsely cratered northern lowlands. Ninety-three percent of craters larger than 30 kilometers in diameter are found south of this line, including the massive Argyre, Hellas, and Isidis basins—ancient scars from collisions with asteroids and clues to the death of a world.

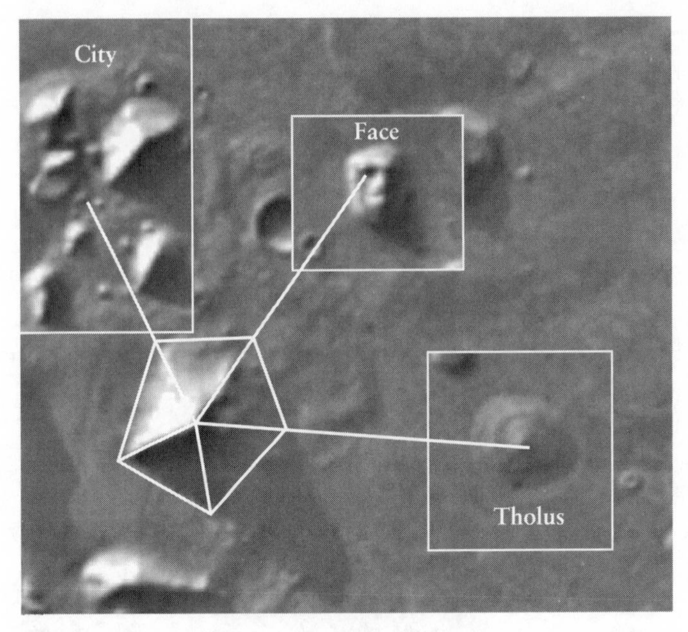

The D&M Pyramid is seemingly aligned to other anomalous features of the Cydonia Mensa—the City Center, the "tear" on the Face, and the apex of the Tholus.

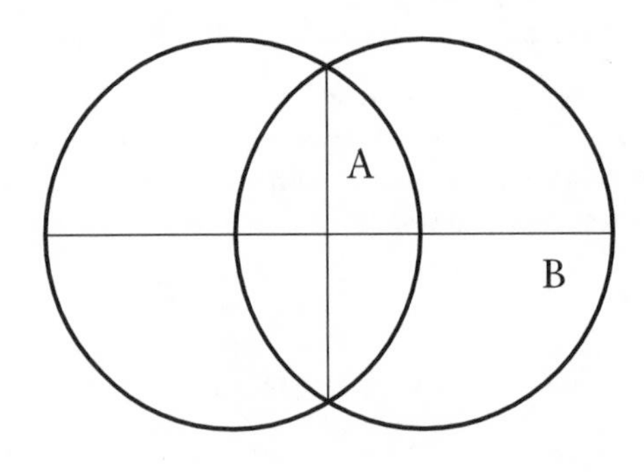

The *vesica piscis,* formed by the overlapping of 2 circles, is an important geometric form in the tradition of sacred geometry, yielding many mathematical constants as well as the golden section, the *phi* ratio, which is formed by the ratio of lengths A to B, roughly 3:5. The *phi* ratio was widely used in ancient terrestrial architecture, and identical geometric constants are found repeated in the measurements of the Cydonian anomalies.

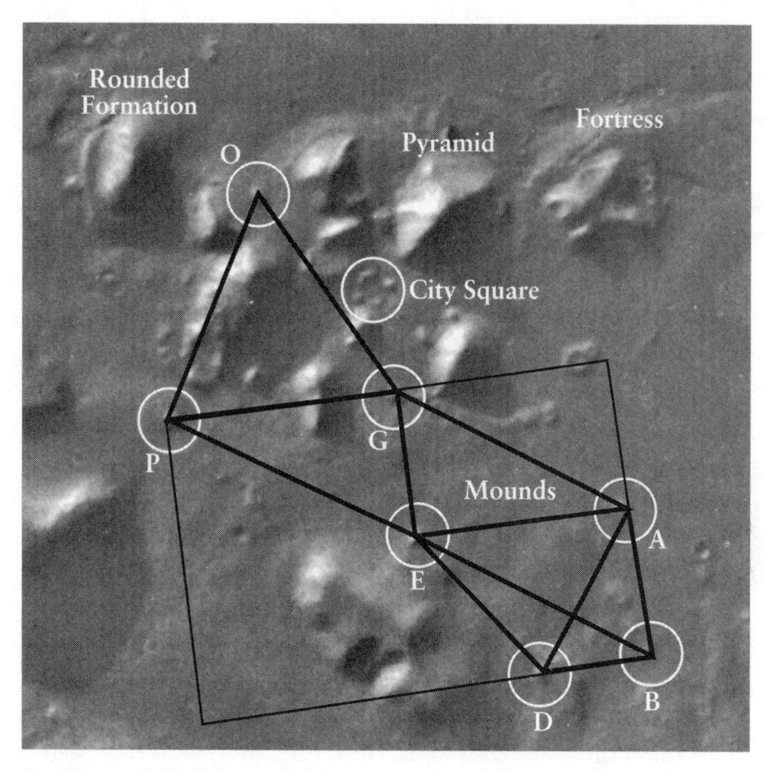

Dr. Horace Crater's analysis of the layout of "mounds" within the City area reveals an alignment that is unlikely to have occurred naturally.

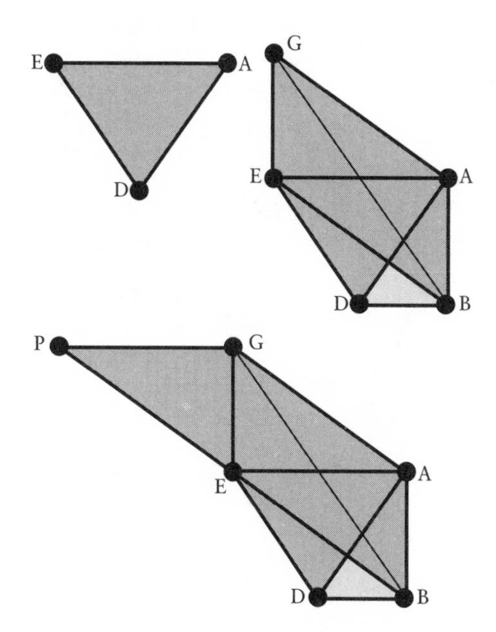

The alignments of mounds EAD, GABDE, and GAB-DEP show a highly unnatural repetition of basic triangles. Is this the work of nature or intelligence?

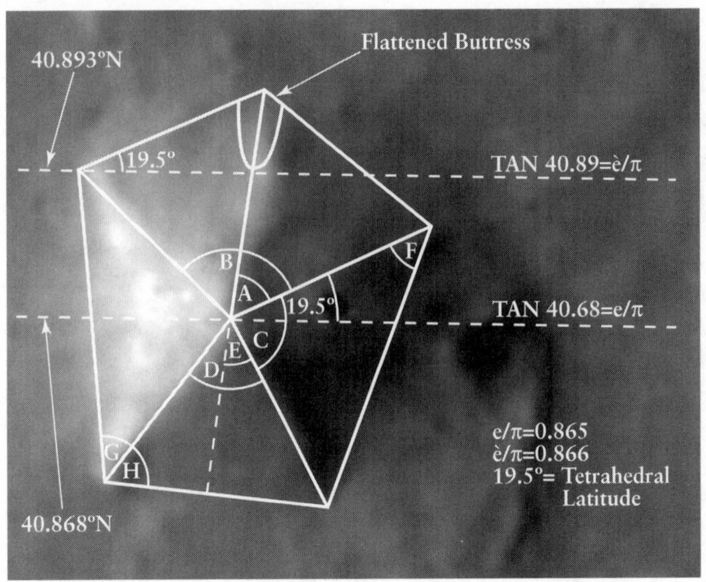

Radians

Radian A = $\pi/3$
Radian B = $2\pi/3$
Radian D = $e/\sqrt{5}$
Radian F = $\grave{e}/\pi$

Angle Ratios

C/A = $\sqrt{2}$
B/D, C/F = $\sqrt{3}$
A/D = e/π
C/D, A/F, H/G = $e/\sqrt{5}$
B/C, D/F = $\pi/\sqrt{5}$

Trigonometric Functions

TAN A, TAN B = $\sqrt{3}$
SIN A, SIN B = $\grave{e}/\pi$
TAN F = π/e
COS E = $\sqrt{5}/e$
SIN G = $\sqrt{5}/\pi$

Erol Torun's reconstructed model of the D&M Pyramid yields unique mathematical constants, including those found in terrestrial traditions of sacred geometry, as well as the tetrahedral angle of 19.5 degrees.

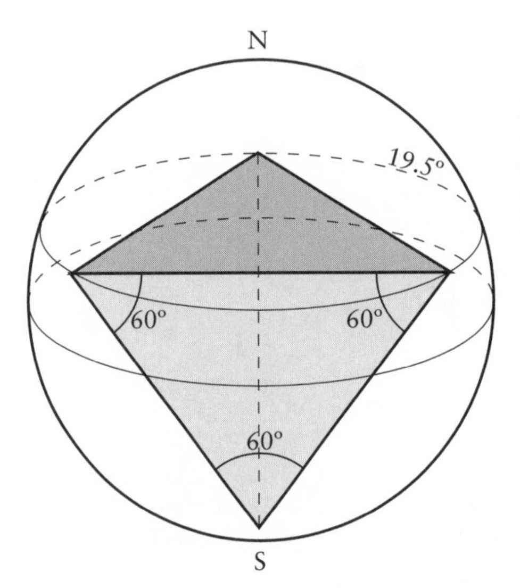

Circumscribed tetrahedron: if a tetrahedron, the simplest of the platonic solids, is placed within a rotating sphere with one apex at the north or south pole, the other three apexes will lie at exactly 19.5 degrees from the equator. This tetrahedral angle of 19.5 degrees occurs with unnatural frequency in the measurements of the Cydonian anomalies. Is this evidence of a lost mathematical message?

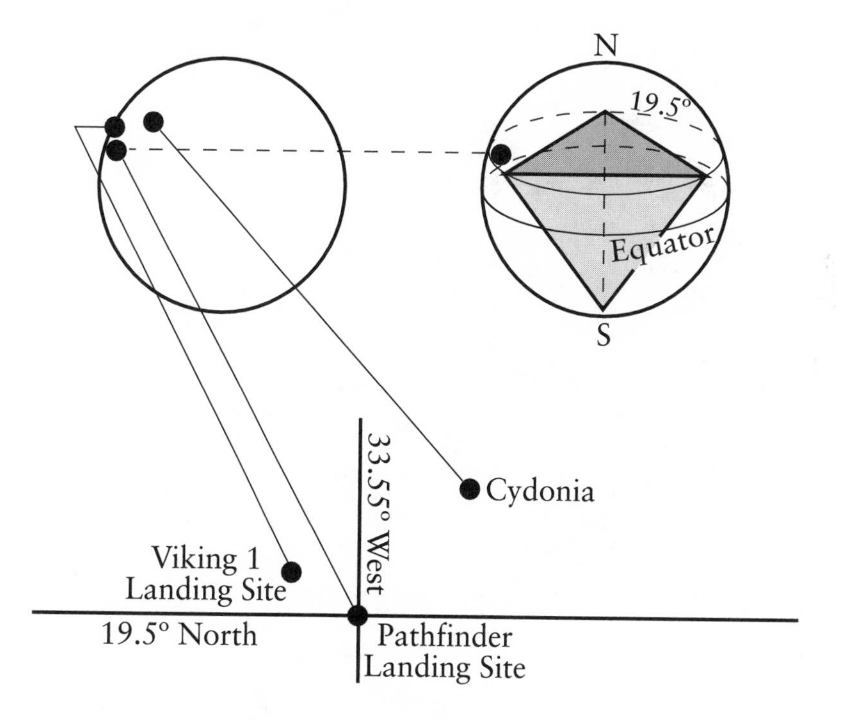

The landing spot of the tetrahedral-shaped *Mars Pathfinder,* coincidentally, lies at roughly 19.5 degrees (the tetrahedral angle) north of the Martian equator.

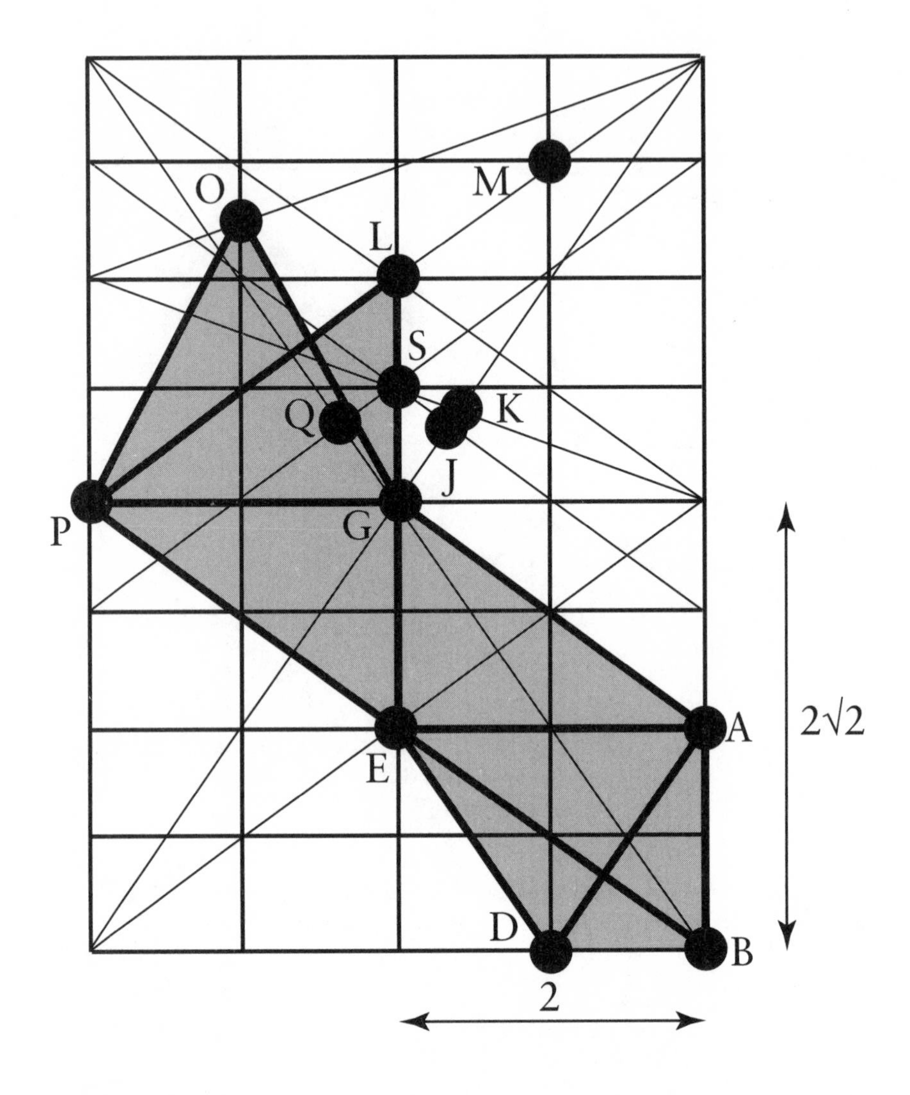

Professor Stanley McDaniel's analysis of the Cydonian mound configuration reveals that all the mounds can be fitted onto a grid based on the square root of 2, a framework also used in ancient terrestrial sacred architecture.

According to researchers Richard Hoagland and Erol Torun, major alignments between the Cydonian anomalies reveal an underlying coherence based on the tetrahedral angle of 19.5 degrees and the polar diameter of Mars.

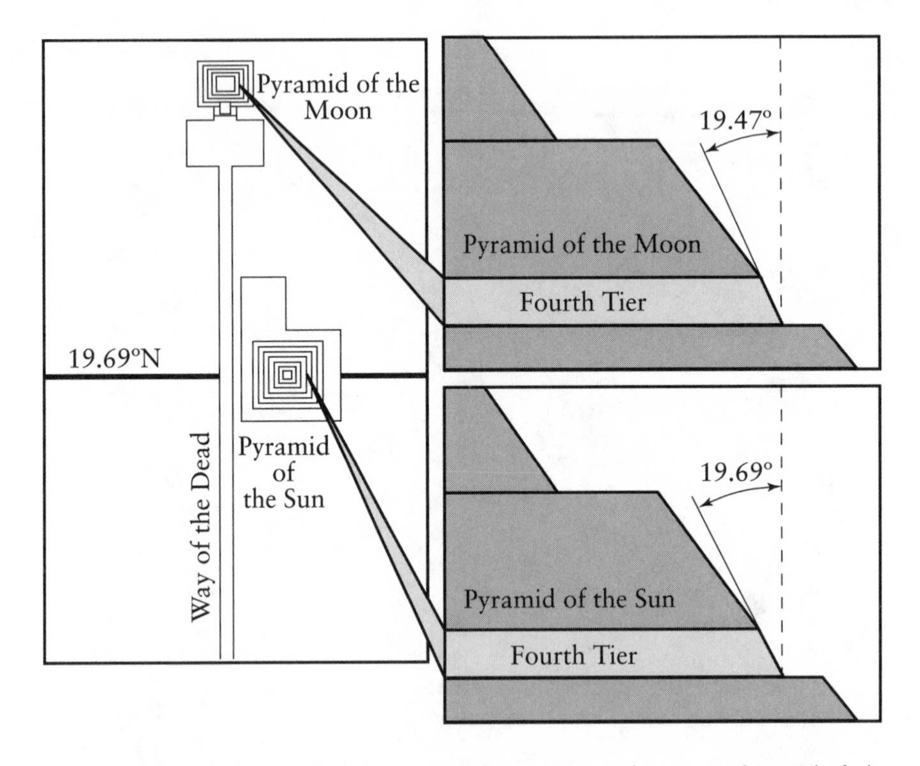

The Pyramids of the Sun and Moon at Teotihuacan contain references to the tetrahedral constant of 19.5 degrees in both the measurement of the angles of the fourth pyramid tiers and in their geographical location on Earth—which coincidentally mirrors the self-referencing of the D&M Pyramid on Mars.

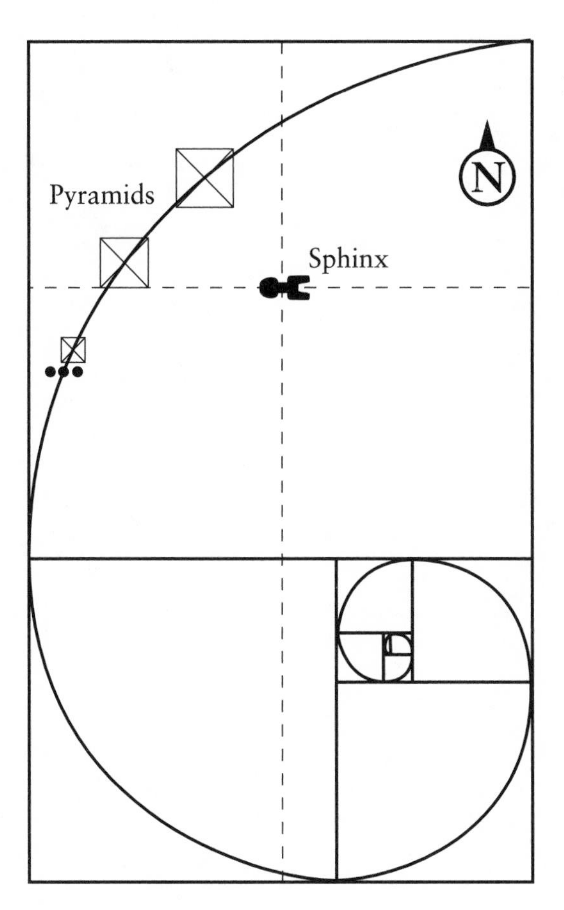

According to researcher Erol Torun, the placement of the Pyramids and Sphinx at Giza are conditioned by the Fibonacci curve (see page 153), based on the ancient sacred proportion of *phi*, the golden section.

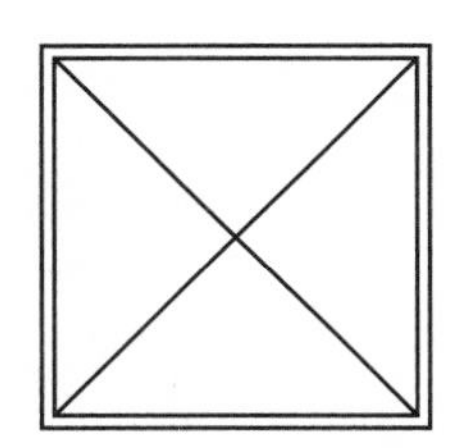

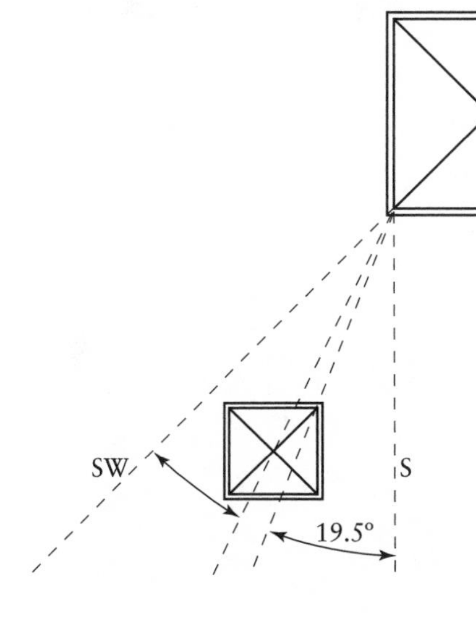

Using calculations based on the work of Egyptologist John Legon, the placement of the smallest of the three pyramids of Giza, the Pyramid of Menkaure, in relation to its neighbors, can be seen to be based on the tetrahedral angle of 19.5 degrees—the same angle mysteriously referred to in the Pyramids of Cydonia.

Endnotes

Part One: The Murdered Planet

1. PARALLEL WORLD

1. *Astronomy Now,* London, 1996, p. 39.
2. Anders Hansson, *Mars and the Development of Life* (Chichester and New York: John Wiley and Sons, 1997), 53.
3. Ibid.
4. Ibid., 52.
5. Ibid., Preface, xiii.
6. *The Sunday Times* (London), 1 December 1996.
7. See discussion in Hansson, *Mars and the Development,* 137–153. See also Arthur C. Clarke, *The Snows of Olympus,* Victor Gollarcz, London, 1994.
8. Ibid.
9. Ibid., 19, 128.
10. Fred Hoyle and Chandra Wickramasinghe, *Lifecloud: The Origin of Life in the Universe* (London and Toronto: J. M. Dent and Sons, 1978).
11. *Encyclopaedia Britannica,* 15th edition, "Solar System."
12. Ibid.
13. Ibid.
14. Mack Gipson Jr. and Victor K. Ablordeppy, *Icarus* 22 (1974), 197–204.
15. Carl Sagan, *Cosmos* (London: Book Club Associates, 1981), 130.
16. Viking project scientist Gerry Soffen cited in Richard C. Hoagland, *The Monuments of Mars* (Berkeley, Calif.), 5.
17. V. DiPietro and G. Molenaar, *Unusual Martian Surface Features* (privately published, 1982), 38; M. Carlotto, *The Martian Enigmas: A Closer Look* (Berkeley, Calif.: North Atlantic Books, 1997), 181.
18. Carlotto, *Martian Enigmas,* 28.
19. Ibid., 28.

20. See Stanley McDaniel, *The McDaniel Report* (Berkeley, Calif.: North Atlantic Books, 1993), 82–84.
21. Hoagland, *Monuments of Mars,* 25.
22. Ibid., 26.
23. Ibid., 27.
24. McDaniel, *Report,* 65–66.
25. Chris O'Kane, telephone conversation with the authors, August 1996.
26. Ibid.
27. DiPietro and Molenaar, *Unusual Martian,* 106–12; Carlotto, *Martian Enigmas,* 88–95; Hoagland, *Monuments of Mars,* 317–21.
28. Carl Sagan, *The Demon Haunted World,* Headline, London, 1996, 56.
29. Ibid., 56.

2. IS THERE LIFE ON MARS?

1. R. S. Richardson and C. Bonestall, *Mars* (London: George Allen and Unwin, 1965), 3.
2. *Encyclopaedia Britannica,* 15th edition, "Solar System."
3. Ibid.
4. *Times* (London), 11 November 1996.
5. Ibid.
6. Ibid.
7. *Times* (London), 8 June 1997.
8. *Guinness Book of Astronomy,* 62ff.
9. *Newsweek,* 23 September 1996, 57.
10. Peter Cattermole, *Mars: The Story of the Red Planet* (London and New York: Chapman and Hall, 1992), 37.
11. *Times* (London), 13 October 1996.
12. Tim Radford, *London Review of Books,* 3 July 1997, 16.
13. Ibid.
14. Ibid.
15. *Times* (London), 8 August 1996.
16. *The Guardian* (London), 1 June 1995.
17. *Newsweek,* 23 September 1996, 57.
18. Radford, *London Review of Books,* 16.
19. Hansson, *Mars and the Development,* 45.

20. Paul Davis, *The Guardian* (London), 1 June 1995.
21. Quoted in *Quest for Knowledge* (Chester), October 1996, 6.
22. *Daily Telegraph* (London), 24 May 1997.
23. *The Sunday Times* (London), 3 November 1996.
24. *Sydney Morning Herald* (Australia), 26 December 1996.
25. *The Sunday Times* (London), March 1997.
26. *Times* (London), 9 June 1997.
27. Ibid.
28. Hansson, *Mars and the Development,* xiii.
29. *Daily Mail* (London), 1 November 1996.
30. *Daily Mail* (London), 8 August 1996.
31. *Times* (London), 8 August 1996.
32. National Academy of Sciences briefing to Vice President Al Gore, 11 December 1996.
33. *Daily Mail* (London), 8 August 1996.
34. *Times* (London), 8 August 1996.
35. *Times* (London), 9 June 1997.
36. Daniel Goldin, quoted in *Spaceflight* 38 (October 1996): 328.
37. *Hieronymous and Co. Newsletter* 1:8–10, 2, Owens Mills, Md.
38. Ibid.
39. Ibid.
40. *New Scientist,* 17 August 1996; *Times* (London), 8 August 1996.
41. Ibid.
42. Ibid.
43. *Astronomy Now,* October 1996, 39–42.
44. "Mars Dossier," *Focus* (1996), 90.
45. *Spaceflight* 38 (October 1996): 327; *Times* (London), 9 August 1996.
46. *Writing in Spaceflight* 38 (October 1996): 328.
47. *The Sunday Times* (London), March 1997.
48. Ibid.
49. Bartholemew Nagy, quoted in *Hieronymous and Co. Newsletter,* 1:8–10, 1.
50. Nagy, quoted in *Nature,* 20 July 1989.
51. "Mars as the Parent Body of the CI Carbonacious Chondrites," *Geophysical Research Letters,* 1 May 1996. Cited in *Hieronymous and Co. Newsletter,* 6.
52. *Hieronymous and Co. Newsletter,* 4.

53. Ibid., 1.
54. Ibid.
55. *Daily Mail* (London), 30 August 1996.
56. *Spaceflight* 38 (October 1996): 328.
57. Ibid.
58. *Hieronymous and Co. Newsletter*, 5.

3. THE MOTHER OF LIFE

1. Miller and Orgel, quoted in Hansson, *Mars and Development*, 38.
2. Ibid., 77–78.
3. Ibid., 37.
4. Ibid.
5. Percival Moore, quoted in Patrick Moore, *New Guide to the Planets*, (London: Sidgwick and Jackson, 1993), 99–100.
6. Lowell in 1894 was the first astronomer to discuss the wave of darkening in detail.
7. *Encyclopaedia Britannica*, 15th edition, "Solar System."
8. Cattermole, *Mars*, 192; *Encyclopaedia Britannica*, 15th edition, "Solar System."
9. *Encyclopaedia Britannica*, 15th edition, "Solar System."
10. Cattermole, *Mars*, 161.
11. Ibid., 161.
12. *Encyclopaedia Britannica*, 15th edition, "Solar System."
13. Ibid.
14. Cattermole, *Mars*, 23–24.
15. Ibid., 91–94.
16. Kim Stanley Robinson, *Green Mars*, quoted in Clarke, *Snows of Olympus*, 55.
17. Cattermole, *Mars*, 104.
18. Ibid., 23, 72.
19. Ibid., 72.
20. Ibid., 23–24; Murray, Malin, Ronald Greely, *Earthlike Planets* (San Francisco: W. H. Freeman, 1981), 297.
21. Cattermole, *Mars*, 30; *Encyclopaedia Britannica*, 15th edition, "Solar System."

22. Cattermole, *Mars,* 30.
23. Ibid., 134.
24. Ibid., 32.
25. Ibid., 22.
26. Ibid.
27. Ibid., 22, 72.
28. Ibid., 22, 27.
29. Donald W. Patten and Samuel L. Windsor, *The Scars of Mars* (Seattle: Pacific Meridien Publishing, 1996), 12; Cattermole, *Mars,* 27.
30. Patten and Windsor, *The Scars of Mars,* 12.
31. Ronald Greely, *Planetary Landscapes* (New York: Chapman and Hall, 1994), 155.
32. Giuseppe Filotto, *The Face on Mars,* 150. See also Cattermole, *Mars,* 25.
33. Filotto, *Face on Mars* (Gardenview, South Africa: Exact Print, 1995), 150.
34. Cattermole, *Mars,* 60.
35. Greely, *Planetary Landscapes,* 175.
36. Melosh and Vickery quoted in John and Mary Gribbin, *Fire on Earth: In Search of the Doomsday Asteroid* (London and New York: Simon and Schuster, 1996), 77.
37. Ibid., 76.
38. Ibid., 79.
39. Hansson, *Mars and the Development,* 68ff.
40. *Scientific American,* November 1996.
41. DiPietro and Molenaar, *Unusual Martian,* 60ff.
42. Ibid.
43. Carr, et al., *An Exobiological Strategy for Mars Exploration,* NASA, January 1995.
44. Ibid., 8–9.
45. Cattermole, *Mars,* 32.
46. *Scientific American,* November 1996.
47. Victor Baker and Daniel Milton, "Erosion by Catastrophic Floods on Mars and Earth," *Icarus 23* (1974): 27–41.
48. Cattermole, *Mars,* 198.
49. *Scientific American,* November 1996.
50. *Charleston Gazette,* 8 July 1997.
51. Ibid.

52. Cattermole, *Mars,* 198; Murray, Malin, Greely, *Earthlike Planets,* 277, 286.
53. Hansson, *Mars and the Development,* 41.
54. Ibid.
55. Ibid.
56. See chapter 2.
57. Cattermole, *Mars,* 130.
58. *Astronomy Now,* October 1996, 45–46.
59. Greely, *Planetary Landscapes,* 185.
60. Cattermole, *Mars,* 198; Greely, *Planetary Landscapes,* 185.
61. Cydonia coordinates from Hoagland, *Monuments of Mars,* 16.
62. *Hieronymous and Co. Newsletter,* 14, 16.

4. THE JANUS PLANET

1. For example, see Gribbin, *Fire on Earth,* 74–75.
2. *Encyclopaedia Britannica,* 15th edition, "Solar System." "Mars moves around the Sun at a mean distance approximately 1.52 times that of the Earth from the Sun. At closest approach Mars is 206,600,000 kilometers from the Sun and 249,200,000 kilometers at its furthest distance. Mars completes an orbit in roughly the time Earth completes two, so spends most of its year far from Earth in directions that are near to the Sun." Closest approach of Mars to Earth: 56,000,000 kilometers. Farthest from Earth: 400,000,000 kilometers.
3. Cattermole, *Mars,* 191.
4. Carr, et al., *Exobiological Survey,* 233–34.
5. William K. Hartmann, "Cratering in the Solar System," *Scientific American,* January 1977, 97.
6. George E. McGill and Steven W. Squires, "Origin of the Martian Crustal Dichotomy: Evaluating Hypotheses," *Icarus* 93 (1991): 386.
7. Ibid., Cattermole, *Mars,* 191.
8. Carr, et al., *Exobiological Survey,* 233–34.
9. Hartmann, "Cratering," 89; Arvidson, Goettel, et al., "A Post-*Viking* View of Martian Geologic Evolution," *Reviews of Geophysics and Space Physics* 18 (August 1980): 575.

10. McGill and Squires, "Origin," 391.
11. Hartmann, "Cratering," 97.
12. L. A. Soderblom, C. D. Condit, et al., "Martian Planetwide Crater Distributions: Implications for Geologic History and Surface Processes," *Icarus* 22 (1974): 240.
13. So far as we are aware the first investigators to give serious consideration to this possibility are Patten and Windsor; see their *Scars of Mars.*
14. D. S. Allan and J. B. Delair, *When the Earth Nearly Died: Compelling Evidence of a Catastrophic World Change, 9500 B.C.* (Bath, England: Gateway Books, 1995), 230.
15. Patten and Windsor, *Scars of Mars,* 18–19.
16. Ibid. See also Patten and Windsor, *The Recent Organization of the Solar System* (Seattle: Pacific Meridien Publishing, 1997).
17. *Icarus* 36 (1978): 51–74.
18. Ibid., 51.
19. Greely, *Planetary Landscapes,* 155.
20. Patten and Windsor, *Scars of Mars,* 19–21.
21. Ibid.
22. Ibid.
23. Ibid., 30–31.
24. Cattermole, *Mars,* 56–58.
25. See part 4.
26. *Mail on Sunday Review* (London), 12 June 1994, 43.
27. Gribbin, *Fire on Earth,* 44; *Encyclopaedia Britannica,* 15th edition, "Tunguska event."
28. Gribbin, *Fire on Earth,* 45; *Mail Review,* 43.
29. Gribbin, *Fire on Earth,* 47–48.
30. Ibid., 30ff.
31. Ibid., 11–12.
32. Ibid., 1, 12.
33. *Encyclopaedia Britannica,* 15th edition, "Solar System."
34. See part 4.
35. Gribbin, *Fire on Earth,* 32; Hartmann, "Crater," 86.
36. Gribbin, *Fire on Earth,* 32.
37. Patten and Windsor, *Scars of Mars,* 31.
38. Ibid.

39. Ibid., 37; Cattermole, *Mars,* 30.
40. Cattermole, *Mars,* 142.
41. Gribbin, *Fire on Earth,* 78.
42. Hartmann, "Crater," 97.
43. Allen and Delair, *When Earth Nearly Died,* 205.
44. Ibid.
45. Hartmann and Larson, "Angular Momenta of Planetary Bodies," *Icarus* 7 (1967): 258; see also Fish, "Angular Momenta of the Planets," *Icarus* 7 (1967): 251ff.
46. Allen and Delair, *When Earth Nearly Died,* 205.
47. "Large-Scale Variations in the Obliquity of Mars," *Science* 181: 4096, 260ff.
48. Jihad Touma and Jack L. Wisdom, *Scientific American,* November 1996. Emphasis added.
49. "Large-Scale Variations," 205–6. See also Cattermole, *Mars,* 9. Mars has a weak magnetic field—only 0.03 percent that of Earth.
50. Peter H. Schultz, "Polar Wandering on Mars," *Icarus* 73 (1988): 91–141.
51. Hartmann, "Crater," 89.
52. Ibid.
53. Patten and Windsor, *Scars of Mars,* 22.
54. Ibid., 69.
55. Allen and Delair, *When Earth Nearly Died,* 210.
56. Ibid.
57. Victor Clube and William Napier, *The Cosmic Serpent* (New York: Universe Books, 1982); and *The Cosmic Winter* (Oxford, England: Basil Blackwell, 1990).
58. See discussion in Hartmann, "Crater," 89.
59. Cattermole, *Mars,* 175.
60. Filotto, *Face on Mars,* 151.
61. The possibility of such an illusion was specifically recognized by Soderblom, Condit, et al., in *Icarus* 22 (1974): 234, where they observed that curious characteristics of the Martian dichotomy "create the impression that terrains on Mars are either ancient, dating back to the early phase of Martian history, or extremely young, perhaps developed in the latest stages in Martian evolution."

62. Graham Hancock, *Fingerprints of the Gods.*
63. Ibid.
64. *Orion Mystery, Fingerprints of the Gods, The Message of the Sphinx, Heaven's Mirror.*
65. Ibid.
66. Hancock and Bauval, *The Message of the Sphinx.*
67. See part 3.

Part Two: The Mystery of Cydonia

5. CLOSE ENCOUNTER

1. W. Sheehan, *The Planet Mars* (Tucson, Ariz.: University of Arizona Press, 1996), 75.
2. Ibid., 104.
3. Sagan, *Cosmos,* 127.
4. Sheehan, *Planet Mars,* 104.
5. Percival Lowell, address to the Boston Scientific Society, 22 May 1894, quoted in Sheehan, *Planet Mars,* 104.
6. Sheehan, *Planet Mars,* 128.
7. See Richard Noll, "The Jung Cult," chap. 4 (London: Fontana, 1996).
8. Camille Flammarion, *La Planete Mars,* vol. 1, 586.
9. Carl G. Jung, Collected Works, *Psychiatric Studies* vol. 1 (London: Routledge & Kegan Paul, 1957), 34.
10. F. Sarler, "A Sunday Afternoon on Mars," *Sunday Times Magazine* (London), August 1997.
11. P. Moore, *Mission to the Planets* (London: Cassel, 1995), 54.
12. Sagan, *Cosmos,* 134–35.
13. Ibid., 132; Hurtak and Crowley, *The Face on Mars* (Adelaide, Australia: Sun Books, 1986), 2.
14. Sagan, *Cosmos,* 132.
15. Hurtak and Crowley, *The Face on Mars,* 1; Sheehan, *Planet Mars,* 162.
16. Hurtak and Crowley, *The Face on Mars,* 1.
17. Ibid., 125.
18. Sheehan, *Planet Mars,* 164.
19. Moore, *Mission to the Planets,* 125.
20. Sheehan, *Planet Mars,* 164.

21. Moore, *Mission to the Planets,* 125.
22. Sheehan, *Planet Mars,* 165–68.
23. Moore, *Mission to the Planets,* 57.

6. A MILLION TO ONE

1. H. G. Wells, *The War of the Worlds* (London: Pan, 1983), 13–14.
2. *Mars Global Surveyor*'s resolution is 1.4 meters per pixel.
3. Percival Lowell, address to the Boston Scientific Society, 22 May 1894, quoted in Sheehan, *Planet Mars,* 104.
4. Sheehan, *Planet Mars,* 171.
5. Hurtak and Crowley, *The Face on Mars,* 35.
6. Ibid., 36.

7. THE *VIKING* ENIGMA

1. "The plain of gold": named after its coloration.
2. Conversation with authors, July 1997, at Caltech, Pasadena, California.
3. *Cosmos,* 140.
4. Ibid.
5. According to Gerry Soffen, a Viking project scientist.
6. Press release P-17384. (Source: NASA/Internet).
7. *Cosmos,* 140.
8. *The Face on Mars,* 68.

8. JESUS IN A TORTILLA

1. Ares Vallis means simply "Mars Valley."
2. NASA press release (http://nssdc.gsfc.nasa.gov/planetary/text/marsob.txt), 21 August 1993.
3. B. Rux, *Architects of the Underworld* (Berkeley, Calif.: Frog Ltd., 1996), 245.
4. Richard Grossinger, Foreword to Hoagland, *Monuments of Mars,* xxxiii.
5. McDaniel, *Report,* xvi.
6. Ibid.
7. Ibid., 23.

8. Rux, *Architects of the Underworld,* 241–44. The image can be found at the website of the Academy for Future Science (AFFS): affs@affs.org.

9. FACE STARING BACK

1. DiPietro, and Molenaar, *Unusual Martian,* 15. "A fantastic adventure was just beginning."
2. Carlotto, *Martian Enigmas,* 20.
3. DiPietro and Molenaar, *Unusual Martian,* 23.
4. Carlotto, *Martian Enigmas,* 20.
5. Ibid., 18.
6. DiPietro and Molenaar, *Unusual Martian,* 27.
7. Ibid., 38.
8. Jim Channon, quoted in Hoagland, *Monuments of Mars,* 167–68.
9. For face recognition as an "innate releasing mechanism," see A. Stevens, *Archetype: A Natural History of the Self* (London: Routledge, 1992), 57.
10. R. Spitz, "The Smiling Response," in *Genetic Psychology Monographs* 34: 57–125.

10. OZYMANDIAS

1. Mark Carlotto, conversation with the authors, Manchester, England, December 1997.
2. Carlotto, *Martian Enigmas,* 40; and Mark Carlotto, "Digital Imagery Analysis of Unusual Martian Surface Features," *Applied Optics* 27, 15 May 1988.
3. Ibid., 5.
4. McDaniel, *Report,* 48.
5. Carlotto, conversation with authors, December 1997.
6. Carlotto, *Martian Enigmas,* 287.

11. COMPANIONS OF THE FACE

1. Hoagland, *Monuments of Mars,* 16.
2. Ibid., 267.

3. McDaniel, *Report,* 70.
4. Arden Albee, interview with the authors, Cal Tech, Pasadena, California, July 1997.

12. THE PHILOSOPHERS' STONE

1. Sagan, *Cosmos,* 321.
2. Ibid., 324–25.
3. R. Pirsig, *Lila: An Inquiry into Morals* (London: Black Swan, 1992), 392–93.
4. Sagan, *Cosmos,* 134.
5. Hoagland, *Monuments of Mars,* 325.
6. Erol Torun, "The Geomorphology and Geometry of the D&M Pyramid," unpublished paper; available through Compuserve Issues forum, section 10, filename PYRAMI.RSH.
7. Ibid.
8. Ibid.
9. That is, with the orbiter camera angle corrected, so the object is not seen on a slope.
10. David Wood and Dan Campbell, *Genset* (Sudbury on Thames, England: Bellvue Books, 1995), 61.
11. H. E. Huntley, *The Divine Proportion* (New York: Dover Publications, 1970), 24.
12. Wood and Campbell, *Genset,* 61.
13. J. Michell, *The New View Over Atlantis* (London: Thames and Hudson, 1983), 157–59.
14. McDaniel, *Report,* 86.
15. J. Michell, *The New View Over Atlantis,* 157–59.
16. DiPietro and Molenaar, *Unusual Martian,* 38.
17. Phi is 1.6180339885 . . . calculated by adding one to the square root of five and dividing the result by two. See P. Tompkins, *Mysteries of the Mexican Pyramids* (London: Thames and Hudson, 1976), 262.
18. Hoagland, *Monuments of Mars,* 151–52.
19. McDaniel, *Report,* 85.
20. Ibid., 86.
21. J. and C. Matthews, *The Western Way: The Hermetic Tradition* (London: Penguin Arkana, 1988), 199.

22. See collected works of C. G. Jung, *Psychology and Alchemy: Alchemical Studies and Mysterium Conjunctionis.*
23. Rosarium, Art. aurif., II, 237 in Jung, *Psychology and Alchemy.*
24. Jung, *Psychology and Alchemy,* 178.
25. Torun, "Geomorphology and Geometry of D&M Pyramid."
26. Ibid.

13. COINCIDENCES

1. McDaniel, *Report,* 88.
2. Hoagland, *Monuments of Mars,* 326, note 4, Appendix 2; Carlotto, *Martian Enigmas,* 178.
3. Hoagland, *Monuments of Mars,* 351–52.
4. Used as letter *t* in H. Crater and S. McDaniel, *Mound Configurations on the Martian Cydonia Plain: A Geometric and Probablistic Analysis,* privately published, 1995. Camp Ares Ltd., U.K.
5. "The Martian Mysteries," *Quest* 1, (1997): 35.
6. Hoagland, *Monuments of Mars,* fig. 10, and McDaniel, *Report,* 115–16.
7. Crater and McDaniel, *Mound Configurations,* 2.
8. Ibid., 2.
9. Ibid.
10. Ibid., 4.
11. Ibid., 7.
12. "Martian Mysteries," 35.
13. Crater and McDaniel, *Mound Configurations,* 9.
14. Ibid., Appendix C.
15. Ibid., 9.
16. Hoagland, *Monuments of Mars,* 352 and fig. 30.
17. Ibid., 352.
18. Ibid., 469. Also J. McDowell, "*Mars Pathfinder* Update," *Sky and Telescope* 88, (December 1994).

Part Three: Hidden Things

14. DISINFORMATION

1. B. Rux, *Architects of the Underworld* (Berkeley, Calif.: Frog Ltd., 1996), 246.

2. U.S. House, *Report on the Committee on Science and Astronautics,* 87th Cong. 1st sess., 242; *Proposed Studies on the Implications of Peaceful Space Activities for Human Affairs,* prepared for NASA by the Brookings Institute and delivered to the Committee of the Whole House of the State of the Union, 18 April 1961.
3. *Architects of the Underworld,* 246.
4. Hoagland, *Monuments of Mars,* 409.
5. Ibid., 410.
6. Brookings Institute, *Implications of Peaceful Space.*
7. CIA, *Report of Meetings of Scientific Advisory Panel on Unidentified Flying Objects Convened by Scientific Intelligence,* 14–18 January 1953. This panel was later named the Robertson Panel after its chairman, Dr. H. P. Robertson. Quoted in Victoria Alexander, *The Alexander UFO Religious Crisis Survey* (Las Vegas, Nev.: Bigelow Foundation, 1994).
8. Filotto, *Face on Mars,* 360.
9. *The Sunday Times* (London), 8 June 1997.
10. Ibid.
11. Ibid.
12. *Times* (London), 25 June 1997.
13. Ibid.
14. Ibid.
15. Alexander, *UFO Religious Crisis Survey,* 1.
16. Ibid., 28.
17. Stanley McDaniel, lecture delivered at Quest for Knowledge conference, Harpenden, England, 27 September 1997.
18. Hoagland, *Monuments of Mars,* 206–8, and Carlotto, *Martian Enigmas,* 196.
19. Hoagland, *Monuments of Mars,* 206–8.
20. "I hope that forthcoming American and Russian missions, especially orbiters with high-resolution television cameras, will make a special effort—among hundreds of other scientific questions—to look more closely at the Pyramids and what some call the Face and the City." Sagan, *Demon,* New York: Quoted in Hoagland, *Monuments of Mars,* 471.
21. Malin Space Science Systems Web site (www.msss.com).
22. McDaniel, Quest for Knowledge lecture.
23. Malin Space Science Systems Web site (www.msss.com).

15. CAMERA OBSCURA

1. The girls always claimed that they were trying to reproduce images of real fairies they had seen. See J. Bord, *Fairies: Real Encounters with Little People* (London: Michael O'Mara Books, 1997).
2. Cong. Howard Wolpe, quoted in *Architects of the Underworld,* 246.
3. McDaniel, *Report,* 166–67.
4. McDaniel, Quest for Knowledge lecture.
5. McDaniel, *Report,* 15.
6. Ibid., 23–24.
7. Ibid., 168.
8. See http://nssdc.gsfc.nasa.gov/planetary/text/marsob.txt.
9. G. E. Cunningham, in NASA's JPL Mars Exploration Program's publication *The Martian Chronicle,* no. 1, January 1995.
10. Malin Space Science Systems website (www.msss.com).
11. Cunningham, in *Martian Chronicle,* 4.
12. NASA, press release, in Hoagland, *Monuments of Mars,* 431.
13. Arden Albee, interview with the authors at Cal Tech, Pasadena, California, 19 July 1997.
14. AUFORA news update via CNI News.
15. Ibid.
16. Albee, interview, 19 July 1997.
17. McDaniel, *Report,* 20.
18. Esposito, Roth, and Demeak, *Journal of Spacecraft and Rockets* 28, (Sept.–Oct. 1991).
19. McDaniel, *Report,* 22.
20. Ibid., 22.
21. McDaniel, Quest for Knowledge lecture.

16. CITIES OF THE GODS

1. See Graham and Bauval, *Mysteries of the Mexican Pyramids,* 244–45.
2. J. Michell, *The New View Over Atlantis* (San Francisco: Harper and Row, 1983), 131.
3. Ibid., 131.
4. See discussion in Hancock, *Fingerprints of the Gods.*
5. See Hancock and Faiia, *Heaven's Mirror.*
6. Hancock, *Fingerprints of the Gods.*

7. Pete Tompkins, *Mysteries of the Mexican Pyramids,* 263.
8. Ibid., 279.
9. Ibid., 263.
10. Ibid., 251.
11. McDaniel, *Report,* 142.
12. Hoagland, *Monuments of Mars,* 358.
13. For Avebury, see *The Face on Mars: The Avebury Connection,* VHS video presented by David Percy (London: Aulis Publishing).
14. Hoagland, *Monuments of Mars,* fig 40.
15. Skyglobe 3.6, Klassin Software, Ann Arbor, Michigan, 1993.
16. J. Legon, "A Ground Plan at Giza," *Discussions in Egyptology 10* (1988): 35.
17. Ibid., 33.
18. Ibid., 34–35.
19. Measurement by the authors is based on Legon's measurements of the Giza plateau.

17. THE FEATHERED SERPENT, THE FIRE-BIRD, AND THE STONE

1. Hancock and Bauval, *Mysteries of the Mexican Pyramids,* 266–69.
2. Ibid., 271.
3. Appian Way, *The Riddle of the Earth* (London: Chapman and Hall, 1925), 165.
4. Joseph Campbell, *The Mythic Image* (New York: Princeton-Bollingen Series, 1974), 141.
5. Ibid., 149, quoting H. Jacobi, "Indian Ages of the World," in *Encyclopedia of Religion and Ethics,* ed. James Hastings (New York: 1928), 201.
6. In Hancock, *Fingerprints of the Gods* (London, 1996), 213–14.
7. Ibid., 106–8.
8. E. A. E. Reymond, *The Mythical Origin of the Egyptian Temple* (New York: Manchester University Press, Barnes and Noble, 1969).
9. Ibid., 113.
10. Ibid., 109, 113–14, 127.
11. Ibid., 77.
12. Robert Bauval and Adrian Gilbert, *The Orion Mystery.* (London: Heinemann, 1994), 203ff.

13. Ibid., 300.
14. Ibid., 201.
15. Ibid., 202.
16. Ibid., 202.
17. Ibid., 203.
18. Ibid., 202.
19. Ibid., 203.
20. Ibid., 203.
21. *Encyclopaedia Britannica,* 15th edition.
22. Who was also said to have died and returned to life.
23. Hancock and Bauval, *The Orion Mystery, The Message of the Sphinx, Heaven's Mirror.*
24. Ibid.
25. See discussion in *The Message of the Sphinx.*
26. *The Orion Mystery.*
27. Ibid.
28. *The Message of the Sphinx.*
29. See discussion in *Fingerprints of the Gods.*
30. *The Message of the Sphinx.*
31. Ibid.
32. *The Orion Mystery.*
33. *The Orion Mystery; The Message of the Sphinx.*
34. Ibid.
35. R. O. Faulkner, ed., *The Ancient Egyptian Pyramid Texts* (Oxford University Press, 1969).
36. Ibid.
37. Ibid.
38. Ibid. See also Jane Sellers, *The Death of Gods in Ancient Egypt* (London: Penguin, 1992).
39. *The Message of the Sphinx.*
40. *Fingerprints of the Gods; Heaven's Mirror.*
41. See discussion in *Fingerprints of the Gods.*
42. *The Message of the Sphinx.*
43. Hoagland, *Monuments of Mars,* 287.
44. Ibid., 289.
45. Otto Neugebauer and Richard A. Parker, *Egyptian Astronomical Texts,* Vol. III (London: Lund Humphries, 1960), 179.

46. Ibid.
47. *The Message of the Sphinx.*
48. Alain Danielou, *The Myths and Gods of India* (Rochester, Ver.: Inner Traditions International, 1991), 166.
49. Robert Graves, *The Greek Myths* (London: Penguin, 1960).
50. Quoted in Wood and Campbell, *Genset,* 279.
51. Shakespeare, *Henry VI,* 1.1.
52. Milton, *Paradise Lost.*

Part Four: The Darkness and the Light

18. THE MOON IN JUNE

1. In the Julian calendar used in Gervase's time 18 June. This date converts to 25 June in the Gregorian calendar, which was introduced in 1582 by Pope Gregory XIII and is used to this day.
2. From the chronicle of Gervase of Canterbury, in Clube and Napier, *Cosmic Winter,* 159.
3. Physicist Graeme Waddington, quoted in David H. Levy, *The Quest for Comets: An Explosive Trail of Beauty and Danger* (New York: Oxford University Press, 1995), 132.
4. Clube and Napier, *Cosmic Winter,* 159–60.
5. Ibid.
6. Ibid., 161.
7. Ibid., 161–62.
8. Levy, *Quest for Comets,* 144.
9. Clube and Napier, *Cosmic Winter,* 161–62.
10. Ibid., 162; Levy, *Quest for Comets,* 130.
11. Gerrit L. Verschuur, *Impact: The Threat of Comets and Asteroids* (New York: Oxford University Press, 1996), 10.
12. Levy, *Quest for Comets,* 130.
13. See chapter 4.
14. Fred Hoyle and Chandra Wickramasinghe, *Life on Mars? The Case for a Cosmic Heritage* (Bristol, England: Clinical Press, 1997), 179.
15. Clube and Napier, *Cosmic Serpent,* 140.
16. Ibid.
17. Clube and Napier, *Cosmic Winter,* 156.
18. Ibid., 156.
19. Ibid., 155.

20. H. J. Melosh, *Impact Cratering: A Geologic Process* (New York: Oxford University Press, 1989), 207.
21. Clube and Napier, *Cosmic Winter,* 156.
22. Trevor Palmer, *Catastrophism, Neocatastrophism and Evolution* (Nottingham, England: Society for Interdisciplinary Studies, 1994), 6.
23. Duncan Steel, *Rogue Asteroids and Doomsday Comets: The Search for the Million Megaton Menace that Threatens Life on Earth* (New York: John Wiley and Sons, 1995), 58–59.
24. Clube and Napier, *Cosmic Serpent,* 140.
25. Palmer, *Catastrophism,* 6.
26. Donald W. Cox and James H. Chestek, *Doomsday Asteroid: Can We Survive* (Amherst, N.Y.: Prometheus Books, 1996), 17.
27. Verschuur, *Impact,* 133; M. E. Bailey, S. V. M. Clube, W. M. Napier, *The Origin of Comets* (Oxford and New York: Pergamon Press, 1990), 397–99.
28. Clube and Napier, *Cosmic Winter,* 150.
29. Richard Leaky and Roger Lewin, *The Sixth Extinction: Biodiversity and its Survival* (London: Weidenfeld and Nicholson, 1996), 47.
30. *Science,* 25 July 1997.
31. Dr. Joseph Kirschvink, quoted in Cal Tech press release, 24 July 1997.
32. Schultz, "Polar Wandering on Mars."
33. Hancock, *Fingerprints of the Gods.*
34. Rand and Rose Flem-Ath, *When the Sky Fell* (Toronto, Canada: Stoddart, 1995).
35. William Glen, ed., *The Mass-Extinction Debates: How Science Works in a Crisis* (Stanford, Calif.: Stanford University Press, 1994), 25.
36. Cal Tech press release, 24 July 1997.
37. Walter Alvarez, *T-Rex and the Crater of Doom* (Princeton, N.J.: Princeton University Press, 1997), 15.
38. Ibid., 141; David M. Raup, *The Nemesis Affair: A Story of the Death of Dinosaurs and the Ways of Science* (New York: W. W. Norton, 1986), 158.
39. Verschuur, *Impact,* 7; Raup, *Nemesis,* 49.
40. Verschuur, *Impact,* 7.
41. Raup, *Nemesis,* 158.
42. Luis W. Alvarez, *Science,* June 1980.
43. Fred Hoyle, *Ice* (London: Hutchinson, 1981), 167; Alvarez, *Crater of Doom,* 7.

44. Verschuur, *Impact,* 28.
45. Alvarez, *Crater of Doom,* 15; David Brez-Carlisle, *Dinosaurs, Diamonds and Things from Outer Space: The Great Extinction* (Stanford, Calif.: Stanford University Press, 1995), 102.
46. Alvarez, *Crater of Doom,* 9.
47. Verschuur, *Impact,* 123.
48. Paul J. Thomas, Christopher F. Chyba, Christopher P. McKay, *Comets and the Origin and Evolution of Life* (New York: Springer Verlag, 1997), 225.
49. Alvarez, *Crater of Doom,* 14; Thomas, Chyba, McKay, *Comets and Origin,* 225.
50. Alvarez, *Crater of Doom,* 14.
51. Verschuur, *Impact,* 10; Claude C. Albritton Jr., *Catastrophic Episodes in Earth History* (London and New York: Chapman and Hall, 1989), 109.
52. Brez-Carlisle, *Dinosaurs,* 169–70.

19. SIGNS IN THE SKY

1. David Morrison, quoted in Patricia Barnes-Svarney, *Asteroid: Earth-Destroyer or New Frontier?* (New York and London: Plenum Press), 246.
2. NASA, Fact Sheet on Asteroid and Comet Impacts, 1997, and authors' personal communication with David Morrison, NASA, 3 February 1998.
3. Barnes-Svarney, *Asteroid,* 246ff.
4. Ibid., 247.
5. Ibid., 248; and see *Natural Catastrophes During Bronze Age Civilizations,* Second SIS Cambridge Conference, 11–13 July 1997, 5, 6; Verschuur, *Impact,* 199.
6. Quoted in Steel, *Rogue Asteroids,* 254.
7. Hoyle, *Ice,* 144.
8. Brez-Carlisle, *Dinosaurs,* 169–70.
9. George Foster, *The Meteor Crater Story: Full Dramatic Story of the World's First Proven Meteorite Crater* (Meteor Crater Enterprises, 1993), 10–15; Kathleen Mark, *Meteorite Craters* (Tucson, Ariz.: University of Arizona Press, 1987), 25–39.

10. Barnes-Svarney, *Asteroid,* 157; Verschuur in *Impact* puts the rate of discovery at 3–5 per year (p. 148).
11. John S. Lewis, *Rain of Iron and Ice: The Very Real Threat of Comet and Asteroid Bombardment* (New York: Addison-Wesley Publishing, 1996), 88.
12. Ibid.
13. Ibid.
14. Ibid.
15. Barnes-Svarney, *Asteroid,* 71.
16. Verschuur, *Impact,* 150.
17. Melosh, *Impact Cratering,* 215.
18. Ibid.
19. Ibid., 7.
20. Steel, *Rogue Asteroids,* 91.
21. Quoted in Cox and Chestek, *Doomsday Asteroid,* 30.
22. Steel, *Rogue Asteroids,* 59.
23. Ibid.
24. Ibid., 203.
25. Clube and Napier, *Cosmic Serpent,* 62; Hoyle, *Ice,* 141; Levy, *Quest for Comets,* 149.
26. Clube and Napier, *Cosmic Serpent,* 72.
27. Ibid.
28. Ibid., 72; Hoyle, *Ice,* 141.
29. Levy, *Quest for Comets,* 148.
30. Cox and Chesteck, *Doomsday Asteroid,* 298.
31. Ibid.
32. *Evening Standard* (London), 12 March 1998; *Daily Telegraph* (London), 13 March 1998; *Guardian* (London), 13 March 1998; *Independent* (London), 13 March 1998.
33. Lewis, *Rain of Iron and Ice,* 75.
34. Barnes-Svarney, *Asteroid,* 2; see also Palmer, *Catastrophism,* 6–7.
35. Duncan Steel, quoted in Verschuur, *Impact,* 112.
36. Barnes-Svarney, *Asteroid,* 168.
37. Ibid., 169.
38. Ibid.
39. Ibid.
40. Lewis, *Rain of Iron and Ice,* 86.

41. Ibid., 87.
42. Gribbin, *Fire on Earth,* 58; Verschuur, *Impact,* 33.
43. Verschuur, *Impact,* 33.
44. Lewis, *Rain of Iron and Ice,* 85.
45. Ibid.
46. Ibid., 86.
47. Ibid., 85.
48. Verschuur, *Impact,* 69.
49. Steel, *Rogue Asteroids,* 105.
50. *Encyclopaedia Britannica,* 15th edition, "Solar System."
51. Ibid.
52. Verschuur, *Impact,* 69.
53. Gribbin, *Fire on Earth,* 73.
54. Plato, *Timaeus and Critias* (London: Penguin Classics, 1977), 46.
55. Ibid.
56. David H. Levy, *Impact Jupiter: The Crash of Comet Shoemaker-Levy 9* (New York and London: Plenum Press, 1995), 159.
57. *Encyclopaedia Britannica,* 15th edition, "Jupiter."
58. Moore, *Planets,* 128.
59. Verschuur, *Impact,* 170.
60. Levy, *Impact Jupiter,* 259; Gribbin, *Fire on Earth,* 131.
61. Verschuur, *Impact,* 178; Levy, *Impact Jupiter,* 258–59.
62. Steel, *Rogue Asteroids,* 248.
63. Levy, *Impact Jupiter,* 2.
64. Ibid., 45.
65. Ibid., 48–49.
66. Ibid., 49.
67. Ibid., 158.
68. Ibid.
69. Ibid., 167.
70. Ibid., 170.
71. Ibid.
72. Ibid., 173.
73. Verschuur, *Impact,* 187.
74. Levy, *Impact Jupiter,* 176.
75. Verschuur, *Impact,* 177, 184.
76. Ibid., 178.

77. Levy, *Impact Jupiter,* 210.
78. Caroline Shoemaker, quoted in Ibid., 113.

20. APOCALYPSE NOW

1. See chapter 18.
2. Owen B. Toon, et al., "Environmental Perturbations Caused by the Impacts of Asteroids and Comets," *Reviews of Geophysics* 35 (February 1997): 46, 48–49.
3. Ibid., 47.
4. Hoyle and Wickramasinghe, *Lifecloud,* 107.
5. Clube and Napier, *Cosmic Serpent,* 81.
6. Trevor Palmer, "The Fall and Rise of Catastrophism," lecture delivered at Nottingham Trent University, 25 April 1996, 11.
7. Emilio Spedicato, *Apollo Objects* (Bergamo, Italy: Instituto Universitario di Bergamo, 1990), 17.
8. Clube and Napier, *Cosmic Winter,* 222.
9. Ibid.
10. Ibid., 8.
11. Ibid.
12. Ibid., 8.
13. See chapter 18.
14. Clube and Napier, *Cosmic Serpent,* 99.
15. Clube and Napier, *Cosmic Winter,* 8–9.
16. Palmer, "Catastrophism lecture," 11.
17. Spedicato, *Apollo Objects,* 17.
18. Clube and Napier, *Cosmic Serpent,* 101.
19. Cited in Glen, *Mass-Extinction Debates,* 19.
20. Steel, *Rogue Asteroids,* 57–58.
21. Lewis, *Rain of Iron and Ice,* 205.
22. Steel, *Rogue Asteroids,* 49.
23. Verschuur, *Impact,* 159.
24. Palmer, *Catastrophism,* 6.
25. Levy, *Quest for Comets,* 205.
26. Steel, *Rogue Asteroids,* 49.
27. Jack Hills and Patricia Goda, quoted in Lewis, *Rain of Iron and Ice,* 150.

28. Ibid.
29. Verschuur, *Impact,* 153.
30. Steel, *Rogue Asteroids,* 40.
31. Verschuur, *Impact,* 153.
32. Hills and Goda, quoted in Verschuur, *Impact,* 154.
33. Clube and Napier, *Cosmic Serpent,* 102.
34. Ibid., 102.
35. Don Gault, quoted in Spedicato, *Apollo Objects,* 21.
36. Ibid., 21–22.
37. Ibid., 22.
38. Ibid.
39. Emiliani, Kraus, Shoemaker, quoted in Albritton, *Catastrophic Episodes,* 114–15.
40. Ibid.
41. Clube and Napier, *Cosmic Serpent,* 103.

21. EARTH CROSS

1. See chapter 19.
2. Tom Van Flandern, *Dark Matter, Missing Planets and New Comets: Paradoxes Resolved, Origins Illuminated* (Berkeley, Calif.: North Atlantic Books, 1993), 215–36.
3. Hoyle, *Ice,* 143.
4. *Encyclopaedia Britannica,* 15th edition, "Solar System"; Moore, *Planets,* 119, 123.
5. Steel, *Rogue Asteroids,* 126–27; Thomas, Chyba, McKay, *Comets and Origin,* 216; Clube and Napier, *Cosmic Winter,* 259–60; *Encyclopaedia Britannica,* 15th edition, "Solar System."
6. Verschuur, *Impact,* 44.
7. Palmer, *Catastrophism,* 8; Steel, *Rogue Asteroids,* 127.
8. Steel, *Rogue Asteroids,* 127.
9. Moore, *Planets,* 124.
10. Ibid.
11. Steel, *Rogue Asteroids,* 127.
12. Ibid., 27–28.
13. An example is 1993 HA2, Steel, *Rogue Asteroids,* 127. See also Bailey, Clube, Napier, *Origin of Comets;* they include in this category the

still unnamed minor planets, 3552 (1983 SA), 405 (1979 VA), and 1983 XF.

14. *Encyclopaedia Britannica,* 15th edition, "Solar System."
15. Verschuur, *Impact,* 43.
16. *Encyclopaedia Britannica,* 15th edition, "Solar System." Barnes-Svarney, *Asteroid,* 64.
17. Barnes-Svarney, *Asteroid,* 64.
18. *Encyclopaedia Britannica,* 15th edition, "Solar System." Moore, *Planets,* 115.
19. *Encyclopaedia Britannica,* 15th edition, "Solar System."
20. Cox and Chesteck, *Doomsday,* 325–28.
21. *Encyclopaedia Britannica,* 15th edition, "Solar System."
22. Verschuur, *Impact,* 44.
23. *Nature,* 25 April 1996, 689; Cox and Chesteck, *Doomsday,* 56.
24. *Nature,* 25 April 1996, 689.
25. Cox and Chesteck, *Doomsday,* 57.
26. *Nature,* 25 April 1996, 689.
27. Verschuur, *Impact,* 44–45.
28. Ibid.; Steel, *Rogue Asteroids,* 29; Levy, *Quest for Comets,* 193.
29. *Encyclopaedia Britannica,* 15th edition, "Solar System."
30. Ibid., 578.
31. Palmer, *Catastrophism,* 5.
32. Milton Zysman and Clark Whelton, eds., *Catastrophism 2000* (Toronto: Heretic Press, 1990), 7.
33. Steel, *Rogue,* 29; Clube and Napier, *Cosmic Serpent,* 73.
34. Cox and Chesteck, *Doomsday,* 119.
35. Ibid., and Barnes-Svarney, *Asteroid,* 66–67.
36. Barnes-Svarney, *Asteroid,* 66–67.
37. In 1992. Reported in Cox and Chestek, *Doomsday,* 119.
38. Ibid.
39. Lewis, *Rain of Iron and Ice,* 83.
40. Clube and Napier, *Cosmic Winter,* 152–53; Bailey, Clube, Napier, *Origin of Comets,* 397; Verschuur, *Impact,* 45; Palmer, *Catastrophism,* 6.
41. Agence France Presse, 9 February 1998.
42. Spedicato, *Apollo Objects,* 14.
43. *Encyclopaedia Britannica,* 15th edition, "Solar System."
44. See chapter 19.

45. Brian Marsden, "100 Potentially Hazardous Asteroids," *Harvard-Smithsonian Center for Astrophysics,* 25 September 1997.
46. Clube and Napier, *Cosmic Serpent,* 73.
47. Lewis, *Rain of Iron and Ice,* 81.
48. Palmer, *Catastrophism,* 5; *Encyclopaedia Britannica,* 15th edition, "Solar System."
49. Verschuur, *Impact,* 116.
50. Lewis, *Rain of Iron and Ice,* 83; Cox and Chesteck, *Doomsday,* 314.
51. Col. John M. Urias, et al., "Planetary Defense: Catastrophic Health Insurance for Planet Earth," research paper presented to Airforce 2025, October 1996, chapter 3, 4.
52. Steel, *Rogue Asteroids,* 204.
53. Steel, *Rogue Asteroids,* 204–5.
54. NASA, Fact Sheet, 2.
55. Cited in Steel, *Rogue Asteroids,* 13.
56. NASA, Fact Sheet, 1.
57. See chapter 19.
58. NASA, Fact Sheet, 1.
59. Li Ch'un Feng, quoted in Timothy Ferris, "Is This the End?" *The New Yorker,* 27 January 1997, 46.
60. Ibid.

22. FISHES IN THE SEA

1. Johannes Kepler, in Clube and Napier, *Cosmic Serpent,* 48.
2. Hoyle and Wickramasinghe, *Lifecloud,* 104–5; Palmer, *Catastrophism,* 5.
3. Hoyle and Wickramasinghe, *Lifecloud,* 104–5; *Penguin Dictionary of Astronomy,* 279.
4. Palmer, *Catastrophism,* 5.
5. Along with many technical papers in scholarly journals, Clube and Napier have produced two books for the general public elaborating their theory, *Cosmic Serpent,* and *Cosmic Winter.*
6. Verschuur, *Impact,* 57.
7. Moore, *Planets,* 124.
8. Verschuur, *Impact,* 57.
9. Tom Gehrels, in *Scientific American,* March 1996, 34.
10. Victor Clube, interviewed with Graham Hancock, 13 January 1998.

11. Gribbin, *Fire on Earth,* 125.
12. *Scientific American,* March 1996, 34.
13. See, for example, Ferris, "Is This the End?" *The New Yorker,* 27 January 1997, 47.
14. Brez-Carlisle, *Dinosaurs,* 88–89.
15. Fred Hoyle, *The Origin of the Universe and the Origin of Religion* (R.I. and London: Moyer Bell, Wakefield, 1993), 32.
16. *Encyclopaedia Britannica,* 15th edition, "Solar System."
17. Clube and Napier, *Cosmic Serpent,* 65.
18. Steel, *Rogue Asteroids,* 27–28.
19. Bailey, Clube, Napier, *Origin of Comets,* 397; Palmer, *Catastrophism,* 6.
20. Ibid.
21. Clube and Napier, *Cosmic Serpent,* 75.
22. Bailey, Clube, and Napier, *Origin of Comets,* 395; Clube and Napier, *Cosmic Serpent,* 66.
23. According to Victor Clube, interview with Graham Hancock, 13 January 1998.
24. *Encyclopaedia Britannica,* 15th edition, "Solar System." See also Verschuur, *Impact,* 44; Steel, *Rogue Asteroids,* 126–27; Clube and Napier, *Cosmic Serpent,* 66; Bailey, Clube, Napier, *Origin of Comets,* 395.
25. Palmer, *Catastrophism,* 6; Brez-Carlisle, *Dinosaurs,* 89.
26. Brez-Carlisle, *Dinosaurs,* 88–89.
27. Verschuur, *Impact,* 57.
28. *Encyclopaedia Britannica,* 15th edition, "Solar System." *Penguin Dictionary of Astronomy,* 81.
29. Cox and Chesteck, *Doomsday,* 73; Clube and Napier, *Cosmic Winter,* 111.
30. *Penguin Dictionary of Astronomy,* 178; Clube and Napier, *Cosmic Winter,* 111.
31. *Penguin Dictionary of Astronomy,* 178.
32. Steel, *Rogue Asteroids,* 112; Walter Alvarez et al., *Catastrophes and Evolution: Astronomical Foundations* (Cambridge University Press, 1989), 172–73.
33. Duncan Steel, in Thomas, Chyba, McKay, *Origin of Comets,* 211.
34. Steel, *Rogue Asteroids,* 112; Alvarez, *Catastrophes and Evolution,* 172–73; Cox and Chesteck, *Doomsday,* 122.
35. Brian Marsden, quoted in Levy, *Quest for Comets,* 10.
36. Verschuur, *Impact,* 116.

37. Ibid., 116–17.
38. Ibid., 117.
39. Ibid.
40. Ibid.
41. Ibid., 117.
42. Levy, *Quest for Comets,* 7.
43. Ibid., 8, 11; Verschuur, *Impact,* 117.
44. Levy, *Quest for Comets,* 9.
45. Ibid., 10.
46. Ibid.
47. Ibid.
48. Cited in Ibid, 11.
49. Brian Marsden, cited in Ibid.
50. Ibid., 11; Cox and Chesteck, *Doomsday,* 147.
51. Levy, *Quest for Comets,* 11; Cox and Chesteck, *Doomsday,* 147.
52. Verschuur, *Impact,* 118.
53. Ibid.
54. Dr. Clark Chapman, in Cox and Chesteck, *Doomsday,* 123.
55. Revelation 12:3–4.
56. Cox and Chesteck, *Doomsday,* 74; *Quest for Knowledge,* May 1997, 52.
57. Philip Dauber and Richard Muller, *Three Big Bangs* (New York: Helix Books, 1996), 71.
58. David Morrison, McKay, in Thomas, Chyba, *Comets and Origin,* 254.
59. Ibid.
60. Hoyle and Wickramasinghe, *Lifecloud,* 100.
61. Appian Way, *The Riddle of the Earth* (London: Chapman and Hall Ltd., 1925), 166.
62. Clube and Napier, *Cosmic Serpent,* 63.
63. Levy, *Quest for Comets,* 194.
64. *Penguin Dictionary of Astronomy,* 201.
65. *Encyclopaedia Britannica,* 15th edition, "Solar System." *Catalogue of Cometary Orbits,* 12th ed. (Central Bureau for Astronomical Telegrams). Harvard, 1997.
66. Hoyle and Wickramasinghe, *Life on Mars,* 173–74; Brez-Carlisle, *Dinosaurs,* 4, 107; Verschuur, *Impact,* 7–10.
67. Ignatius Donnelly, *Ragnarok: The Age of Fire and Gravel* (London: Sampson Low, 1888), 85; Dauber and Muller, *Three Big Bangs,* 51.

68. Steel, *Rogue Asteroids,* 126.
69. Dauber and Muller, *Three Big Bangs,* 51.
70. Steel, *Rogue Asteroids,* 126.
71. *The Sunday Times* (London), 27 October 1996.
72. Cox and Chesteck, *Doomsday,* 73.
73. Clube and Napier, *Cosmic Winter,* 138; Donnelly, *Ragnarok,* 409.
74. Donnelly, *Ragnarok,* 409–10.
75. Appian Way, *Riddle,* 163–64.
76. Verschuur, *Impact,* 133; *Penguin Dictionary of Astronomy,* 15–16.
77. Verschuur, *Impact,* 61.
78. Steel, *Rogue Asteroids,* 258.
79. Clube and Napier, *Cosmic Serpent,* 134.
80. See chapters 19 and 20. David Levy, S-L 9's codiscoverer, also found a pair of extremely long-period comets traveling on the same orbits, but one reaching perihelion three months ahead of the other. He submitted his data on these comets to Brian Marsden at the International Astronomical Union, who came up with the following solution: "Some 12,000 years ago, a single comet broke in two as it rounded the Sun. The two parts did not separate right away but stayed together as a double comet until millennia later, and far from the Sun, they began to drift apart." See Levy, *Quest for Comets,* 108.
81. Steel, *Rogue Asteroids,* 257.
82. Ibid.
83. Clube and Napier, *Cosmic Serpent,* 133.
84. Verschuur, *Impact,* 59.
85. Clube and Napier, *Cosmic Serpent,* 133.
86. See chapters 19 and 20.

23. VOYAGER ON THE ABYSS

1. R. O. Faulkner, ed., *The Ancient Egyptian Pyramid Texts* (New York: Oxford University Press, 1969), 70.
2. Ibid., 155.
3. Ibid., 144.
4. *Penguin Dictionary of Astronomy,* 253; *Encyclopaedia Britannica,* 15th edition, "Galaxies," "Milky Way."

5. Ibid., 159.
6. *Penguin Dictionary of Astronomy,* 284; Alvarez, *Catastrophes and Evolution,* 155–59.
7. Clube and Napier, *Cosmic Serpent, Cosmic Winter.*
8. See Hancock and Bauval, *The Orion Mystery.*
9. Ibid.
10. Collins, *Stars and Planets,* 232.
11. Roughly 17,000 to 7000 B.P. See Hancock, *Fingerprints of the Gods.*
12. E. A. Wallis Budge, *The Book of Us Dead* (London and New York: Arkana, 1986), 14–15.
13. *Encyclopaedia Britannica,* 15th edition, "Milky Way."
14. Alvarez, *Catastrophes and Evolution,* 154–55, citing Urasin, 1987.
15. Ibid.
16. Clube, interview with Hancock.
17. Brez-Carlisle, *Dinosaurs,* 114.
18. Clube and Napier, *Cosmic Serpent,* 143.
19. Steel, *Rogue Asteroids,* 98; Alvarez, *Catastrophes and Evolution,* 10, 135.
20. Bailey, Clube, Napier, *Origin of Comets,* 264.
21. Ibid.
22. *Philip's Atlas of the Universe* (London: Reed Consumer Books Ltd., 1996), 175.
23. Gould's Belt, confirmed by Clube, 1 February 1998 by phone; Clube and Napier, *Cosmic Serpent,* 33; Alvarez, *Catastrophes and Evolution,* 157.
24. Walter Scott, ed., *Hermetica* (Boston: Shambhala, 1993), 457.
25. Palmer, *Catastrophism,* 58.
26. Clube, interview with Hancock.
27. Thomas, Chyba, McKay, *Comets and Origin,* 9.
28. Clube and Napier, *Cosmic Serpent,* 36.
29. Ibid., 36, 39.
30. Albritton, *Catastrophic Episodes,* 99.
31. Clube and Napier, *Cosmic Serpent,* 40.
32. Ibid., 215–16.
33. Ibid.
34. Clube and Napier, *Cosmic Winter,* 143.
35. Ibid., 134.
36. Palmer, *Catastrophism,* 5.

37. Clube and Napier, *Cosmic Winter,* 134.
38. Ibid., 134.
39. Clube and Napier, *Cosmic Serpent,* 49.
40. Bailey, Clube, Napier, *Origin of Comets,* 250–51.
41. Palmer, *Catastrophism,* 5; Clube and Napier, *Cosmic Winter,* 134.
42. Hoyle, *Origin of Universe,* 30.
43. Palmer, *Catastrophism,* 5.
44. Clube and Napier, *Cosmic Serpent,* 33–35.
45. Palmer, *Catastrophism,* 57; Albritton, *Catastrophic Episodes,* 102–3.
46. Clube and Napier, *Cosmic Serpent,* 34–35.
47. Ibid., 34–35.
48. Palmer, *Catastrophism,* 58; Albritton, *Catastrophic Episodes,* 370.
49. Clube and Napier, *Cosmic Serpent,* 40.
50. Palmer, *Catastrophism,* 58; Thomas, Chyba, *Comets of Origin,* 229: "The common periodicity in mass extinctions and large impact craters was quickly realized to correspond to the half-period with which the Sun oscillates through the galactic plane, suggesting a plausible source for a wave of comets in a disturbance of the Oort cloud either through stellar encounters or passages through giant molecular clouds."
51. Palmer, *Catastrophism,* 58.
52. Hoyle and Wickramasinghe, *Life on Mars,* 174.
53. Ibid.
54. Ibid.
55. Spedicato, *Apollo Objects,* 10. With regard to the K/T event 65 million years ago, it has been observed that Earth passing through a GMC would pick up large quantities of its chemicals. There is evidence for this at K/T boundary. It has been found that the oxygen content of the atmosphere fell from about 35 percent to 28 percent in the 2 million years prior to the K/T boundary event. See Steel, *Rogue Asteroids,* 99–100.
56. Hancock and Bauval, *Secrets of Mexican Pyramids,* 271.

24. VISITOR FROM THE STARS

1. See Hoyle and Wickramasinghe, *Life on Mars,* 174.
2. Palmer, *Catastrophism,* 58.

3. Alvarez, *Catastrophes and Evolution,* 159.
4. Schwarz and James, in Palmer, *Catastrophism,* 58.
5. Clube and Napier, *Cosmic Serpent,* 215–16.
6. Alvarez, *Catastrophes and Evolution,* 156.
7. Clube and Napier, *Cosmic Winter,* 256.
8. Alvarez, *Catastrophes and Evolution,* 157.
9. Clube and Napier, *Cosmic Winter,* 144, 256.
10. *Vistas in Astronomy* vol. 39 (U.K.: Elsevier Science Ltd., 1996), 684.
11. Clube interview with Hancock.
12. Ibid.
13. Hoyle and Wickramasinghe, *Life on Mars,* 176.
14. Verschuur, *Impact,* 134, 136, 138, 163 (citing Steel); Steel, *Rogue Asteroids,* 135–36, 152; Thomas, Chyba, McKay, *Comets and Origin,* 232; Clube and Napier, *Cosmic Serpent,* 133 and *Cosmic Winter,* 149.
15. Steel, *Rogue Asteroids,* 136.
16. Ibid., 135–36.
17. There is nothing inherently improbable about this. "All that is suggested," says Duncan Steel *(Rogue Asteroids,* 135–36), "is a breakup similar to P/Shoemaker-Levy 9 in 1992, except by a comet at least 100 kilometers across and in an orbit crossing from Jupiter to the Earth."
18. Clube, in Alvarez, *Catastrophes and Evolution,* 88.
19. Hoyle, *Origin of Universe,* 34.
20. Ibid.
21. Ibid.
22. Ibid., 35.
23. Clube, in Alvarez, *Catastrophes and Evolution,* 88.
24. Clube and Napier, *Cosmic Winter,* 145–46: "There is strong evidence that the last giant comet entered an Earth-crossing orbit only a few tens of thousands of years ago, so its asteroidal debris (including its resultant zodiacal cloud) *are in orbit even now.*"
25. Alvarez, *Catastrophes and Evolution,* 105.
26. Clube and Napier, *Cosmic Winter,* 244 and *Cosmic Serpent,* 92.
27. Hoyle, *Origin of Universe,* 26–27, 29.
28. Hancock and Bauval, *Fingerprints of the Gods, The Orion Mystery, The Message of the Sphinx.*
29. Raup, *Nemesis,* 59.
30. For a full discussion of crustal displacement and its implications see Flem-Ath, *When the Sky Fell.*

31. Clube interview with Hancock.
32. Clube and Napier, *Cosmic Serpent,* 92.
33. Thomas, Chyba, McKay, *Comets and Origin,* 232: "Could the most recent Ice Age, or its succession, have been due to changes in the small-body flux over the past 10,000 to 20,000 years . . . ? The inner solar system environment [could be] currently subject to the substantial control of the products of the breakup of a giant comet within the past 20,000 years."
34. Hoyle and Wickramasinghe, *Life on Mars,* 176.
35. Hoyle, *Origin of Universe,* 25–26.
36. Ibid., 25–27; Hoyle and Wickramasinghe, *Life on Mars,* 176–77.
37. Hoyle, *Origin of Universe,* 25.
38. Ibid., 26–27.
39. Hoyle, *Ice,* 28.
40. Hoyle, *Origin of Universe,* 28–29.
41. Verschuur, *Impact,* 104, citing Tollman.
42. Chandra Wickramasinghe, interview with Graham Hancock, 16 January 1998.
43. Hoyle, *Origin of Universe,* 34.
44. Ibid., 31.
45. Hancock and Bauval, *Fingerprints of the Gods,* 254.
46. Ibid., 444–48.

25. BULL OF THE SKY

1. Steel, *Rogue Asteroids,* 36.
2. *Pyramid Texts,* 79.
3. Carlotto, *Martian Enigmas,* 92.
4. Hoyle, *Origin of Universe,* 37, 39, 47; Hoyle and Wickramasinghe, *Life on Mars,* 180.
5. Duncan Steel, in Verschuur, *Impact,* 136.
6. *The Sunday Times* (London), 14 December 1997.
7. Dr. Benny Peiser, *Natural Catastrophes During Bronze Age Civilizations,* Second SIS Cambridge Conference, 11–13 July 1997, 9; *Quest News,* May 1997.
8. *The Sunday Times* (London), 14 December 1997.
9. Ibid.

10. Ibid.; *Times* (London), 8 March 1997; see also Marie Courty, in *Natural Catastrophes During Bronze Age Civilizations,* 7–8.
11. *Times* (London), 8 March 1997; Courty, *Natural Catastrophes,* 7–8.
12. Ibid.
13. *The Sunday Times* (London), 14 December 1997.
14. Courty, *Natural Catastrophes,* 8.
15. Victor Clube, in *Independent* (London), Sunday, 30 March 1997.
16. Clube and Napier, *Cosmic Winter,* 147.
17. Steel, *Rogue Asteroids,* 134.
18. Collins, *Stars and Planets,* 232.
19. Courty, *Natural Catastrophes,* 5.
20. Steel, in Ibid.
21. English Heritage Foundation, telephone interview August 1996.
22. Courty, *Natural Catastrophes,* 5–6.
23. Ibid.
24. *Telegraph* (London), Sunday, 16 November 1997.
25. Courty, *Natural Catastrophes,* 5.
26. Clube and Napier, *Cosmic Serpent,* 146–47.
27. Steel, *Rogue Asteroids,* 133; *Penguin Dictionary of Astronomy,* 84–85.
28. Steel, *Rogue Asteroids,* 133.

26. DARK STAR

1. Clube and Napier, *Cosmic Winter,* 12–13.
2. NASA, Fact Sheet.
3. "Massive Asteroid Will Hit Tomorrow," *Spaceguard UK,* 1 January 1998.
4. Clube and Napier, *Cosmic Winter,* 13.
5. Hoyle, *Origin of Universe,* 62.
6. Ibid.
7. *Penguin Dictionary of Astronomy,* 201–2.
8. Ibid., 202.
9. James M. Robinson, *The Nag Hammadi Library* (New York: Brill, 1988), 352.
10. Ibid., 165.
11. Plato, *Timaeus,* 36.

12. Ibid., 35.
13. Ibid., 35.
14. Emilio Spedicato, *Atlantis and Other Tales* (Bergamo, Italy: University of Bergamo, 1997), 10.
15. *Penguin Dictionary of Astronomy,* 385.
16. Verschuur, *Impact,* 134–35; Steel, *Rogue Asteroids,* 133.
17. Clube and Napier, *Cosmic Winter,* 150–51.
18. Ibid., 149, 150.
19. Ibid., 149.
20. *Penguin Dictionary of Astronomy,* 84–85; Steel, *Rogue Asteroids,* 133.
21. Clube and Napier, *Cosmic Winter,* 152–53.
22. See chapter 25.
23. Clube and Napier, *Cosmic Winter,* 152–53.
24. Ibid., 153.
25. Steel, *Rogue Asteroids,* 124, 134.
26. Ibid.
27. Ibid.
28. Clube and Napier, *Cosmic Winter,* 151.
29. Ibid., 152.
30. Ibid., 219.
31. Ibid.
32. Verschuur, *Impact,* 134–35.
33. The general assumption made by asteroid watchers—if anything, likely to underestimate the total numbers of asteroids—is that only about 10 percent of the total population have so far been discovered.
34. Clube and Napier, *Cosmic Winter,* 151.
35. Clube and Napier, *Cosmic Serpent,* 151; *Origin of Comets,* 398; *Cosmic Winter,* 150.
36. Ibid.; Alvarez, *Catastrophes and Evolution,* 100.
37. Hoyle, *Origin of Universe,* 32–33.
38. *Penguin Dictionary of Astronomy,* 178.
39. Spedicato, *Atlantis,* 10.
40. Ibid.
41. Alvarez, *Catastrophes and Evolution,* 11.
42. Hoyle, *Origin of Universe,* 37; Hoyle and Wickramasinghe, *Life on Mars,* 180.
43. Verschuur, *Impact,* 133; Steel, *Rogue Asteroids,* 133–35.

44. Steel, *Rogue Asteroids,* 134–35.
45. Verschuur, *Impact,* 134; Steel, *Rogue Asteroids,* 182; Dauber and Muller, *Three Big Bangs,* 49–50. Hoyle and Wickramasinghe, *Life on Mars,* 178–79; Palmer, *Catastrophism,* 6; Levy, *Quest for Comets,* 130–32.
46. Clube and Napier, *Cosmic Winter,* 275.
47. Cited in *Hermetica,* 111.

Index